D. Readey
11/94

SEMICONDUCTOR DEVICE PROCESSING

Cover photo courtesy of SEMATECH. SEMATECH is a nonprofit research and development consortium of U.S. semiconductor manufacturers in partnership with the Department of Defense. Based in Austin, Texas, SEMATECH's mission is to create fundamental change in manufacturing technology and the domestic infrastructure to provide U.S. semiconductor companies the continuing capability to be world-class suppliers.

SEMICONDUCTOR DEVICE PROCESSING
Technology Trends in the VLSI Era

Robert N. Castellano
The Information Network
San Francisco, California

Gordon and Breach Science Publishers
USA Switzerland Australia Belgium France Germany
Great Britain India Japan Malaysia Netherlands Russia Singapore

Copyright © 1993 by OPA (Amsterdam) B.V. All rights reserved. Published under license by Gordon and Breach Science Publishers S.A.

Gordon and Breach Science Publishers

5301 Tacony Street, Drawer 330
Philadelphia, Pennsylvania 19137
United States of America

Glinkastrasse 13-15
O-1086 Berlin
Germany

Y-Parc
Chemin de la Sallaz
CH-1400 Yverdon, Switzerland

Post Office Box 90
Reading, Berkshire RG1 8JL
Great Britain

Private Bag 8
Camberwell, Victoria 3124
Australia

3-14-9, Okubo
Shinjuku-ku, Tokyo 169
Japan

58, rue Lhomond
75005 Paris
France

Emmaplein 5
1075 AW Amsterdam
Netherlands

Library of Congress Cataloging-in-Publication Data

Castellano, Robert N.
  Semiconductor device processing : technology trends in the VLSI era / Robert N. Castellano.
    p. cm.
  Includes bibliographical references and index.
  ISBN 2-88124-516-1
  1. Semiconductors—Design and construction. 2. Integrated circuits—Very large scale integration—Design and construction.
  I. Title.
  TK7871.85.C35   1993
  621.381'52—dc20   91-36127
                                                                  CIP

No part of this book may be reproduced or utilized in any form or by any means, electronic or mechanical, including photocopying and recording, or by any information storage or retrieval system, without permission in writing from the publisher. Printed in the United States of America.

CONTENTS

List of Illustrations — xi

Preface — xv

Acknowledgements — xvii

Chapter 1   IC Industry Trends — 1

1.1   Trends in IC Processing Technology — 1
1.2   Trends in Contamination — 3
1.3   Trends in Chemical Usage — 4
1.4   Equipment Trends — 5

Chapter 2   Cleanrooms — 9

2.1   Types of Cleanrooms — 9
        2.1.1   Advantages and disadvantages — 10
2.2   Airborne Particle Monitoring — 13
2.3   Electrostatic Discharge — 14
2.4   Future Trends — 15
        2.4.1   Impact of automation — 15
        2.4.2   Impact of ASIC devices — 21
        2.4.3   Impact of VLSI devices — 22
2.5   Japanese Cleanrooms — 23

Chapter 3   Liquid Chemicals — 25

3.1   Technology Issues — 25
        3.1.1   Acids and solvents — 25
        3.1.2   Resists — 26
3.2   Purity Requirements — 30
        3.2.1   Purification methods — 30
                • *Trends for purity—Trace elements* — 30
        3.2.2   Particulates — 31

		•	Effects on yield	31
		•	Particulate removal techniques	35
		•	Particle monitoring	39
		•	Trends for purity—Particle control	39
3.3	Delivery Alternatives		40
	3.3.1	Bottles		40
		•	Glass and polyethylene	40
		•	New materials	40
	3.3.2	Drums		41
		•	Steel and polyethylene	41
		•	Fluoropolymer	42
		•	Returnable drum economics	43
		•	Suppliers	44
	3.3.3	Minibulk containers		44
	3.3.4	Bulk truck trailers		44
3.4	In-Plant Dispensing and Distribution		44
	3.4.1	Bottles in the cleanroom		44
	3.4.2	On-site packaging		45
	3.4.3	Direct dispensing to point of use		46
3.5	Chemical Reprocessors		46
	3.5.1	Piranha reprocessors		47
		•	Background	47
		•	Suppliers	47
		•	Benefits of piranha reprocessors	48
		•	Piranha reprocessor economics	49
	3.5.2	Hydrofluoric acid reprocessors		51
	3.5.3	Strategic issues and future developments		51
3.6	Supplier Rationalization		52
3.7	Statistical Quality Control		54
	3.7.1	Assay and related items		55
	3.7.2	Trace elements		56
	3.7.3	Particles		57
3.8	Analytical Capabilities		57

Chapter 4	**Gases**			**61**
4.1	Technology Issues		61
4.2	Requirements		62
	4.2.1	Purification alternatives		62
		•	Historical perspective	62
		•	Trends for purity—Consistency	63
	4.2.2	Purity trends		65
	4.2.3	Particulate considerations		70

Contents

vii

	•	*Particle monitoring*	70
	•	*Filtration methods*	71

Chapter 5 Lithography 73

- 5.1 Optical Systems — 73
  - 5.1.1 Proximity/contact aligners — 74
  - 5.1.2 Scanning projection aligners — 76
  - 5.1.3 Step-and-repeat aligners — 78
  - 5.1.4 Mix-and-match — 87
  - 5.1.5 Positive resist enhancement — 88
    - *Contrast enhancement materials* — 89
    - *Anti-reflection coatings (ARC)* — 89
    - *Image reversal* — 90
    - *Post-exposure bake* — 90
    - *Stabilization* — 91
  - 5.1.6 Photoresist materials — 91
- 5.2 Electron Beam Systems — 93
  - 5.2.1 Mask making — 93
  - 5.2.2 Direct write — 97
- 5.3 X-Ray Systems — 101
  - 5.3.1 X-ray sources — 101
  - 5.3.2 Mask making — 109
  - 5.3.3 X-ray steppers — 111
  - 5.3.4 X-ray resists — 113
- 5.4 Ion Beam Systems — 115
  - 5.4.1 Direct write — 115
  - 5.4.2 Ion channel masking — 117
  - 5.4.3 Ion projection — 117
  - 5.4.4 Ion sources — 119
- 5.5 Laser Lithography — 121
- 5.6 New Technologies — 121

Chapter 6 Mask Making, Inspection and Repair 127

- 6.1 Mask Making — 127
  - 6.1.1 Mask blanks — 127
  - 6.1.2 Completed masks — 129
- 6.2 Mask Making Equipment — 134
  - 6.2.1 Optical pattern generators — 134
  - 6.2.2 Electron beam systems — 134
  - 6.2.3 Laser pattern generators — 136

Contents

6.3	Mask Inspection	138
6.4	Mask Repair	138
	6.4.1 Laser repair	140
	6.4.2 Focused ion beam repair	141

Chapter 7   Plasma Etching   145

7.1	Single Wafer Systems	145
	7.1.1 Advantages over batch systems	145
	7.1.2 RIE versus PE	152
	7.1.3 Impact of electron cyclotron resonance	154
	7.1.4 Other sources	158
7.2	Processing Issues	161
	7.2.1 Chlorine versus fluorine processes	161
	7.2.2 Multilevel structures	165
	7.2.3 New metallization materials	166
	7.2.4 GaAs processing	167
7.3	Safety Issues	168
	7.3.1 System design considerations	168
	7.3.2 Gas handling	169
	7.3.3 Reactor cleaning	170

Chapter 8   Thin Film Deposition   173

8.1	Technology Trends	173
	8.1.1 Evaporation	174
	8.1.2 Sputtering	175
	8.1.3 Chemical vapor deposition	178
8.2	Interconnection Deposition	184
	8.2.1 Aluminum and aluminum alloys	185
	8.2.2 Polysilicon and silicides	187
	8.2.3 Refractory metals	190
8.3	Dielectric Deposition	192
	8.3.1 Silicon dioxide	192
	8.3.2 Silicon nitride	194
8.4	Technology Comparisons	196
	8.4.1 Evaporation versus sputtering versus CVD	196
	8.4.2 Single versus batch processing	196

Chapter 9   Laser Processing   203

9.1	Introduction	203
9.2	Laser Reactions	206

9.3	Semiconductor Processing		208
	9.3.1	Laser doping	208
	9.3.2	Wafer inspection	208
	9.3.3	Wafer repair	210
	9.3.4	Circuit failure analysis	211
	9.3.5	Particle detection	212
	9.3.6	End-point detection	214
	9.3.7	Laser-induced deposition	215
	9.3.8	Laser-induced etching	216
	9.3.9	Laser planarization	217
	9.3.10	Laser pantography	218
	9.3.11	Packaging	218
	9.3.12	Laser marking	221
	9.3.13	ASIC processing	223
		• *Laserpath*	223
		• *Lasarray*	224
		• *Lasa Industries*	224
		• *Chip Express*	225

Index   **229**

ILLUSTRATIONS

FIGURES

1.1	Minimum Feature Size for Dynamic RAMs with Time	2
2.1	Cleanroom Tunnel Concept	12
2.2	Contamination Levels during Cleanroom Work Cycles	14
2.3	Class 1 Cleanroom Configuration	16
2.4	SMIF Concept	17
2.5	Contamination Levels with and without SMIF Box with Equipment in Aisles	18
2.6	Contamination Levels with and without SMIF Box with Equipment in Environmental Chamber	18
2.7	Air Cleanliness Measurements with and without SMIF Boxes	19
2.8	WAFEC Concept of Localized Wafer Protection	20
3.1	Relationship between Feature Size and Critical Size of Particles	32
3.2	Relationship between Die Yield, Defect Density and Chip Size	33
3.3	Relationship between Die Yield and Cost per Chip	34
3.4	Particle Count Reduction with Chemical Reprocessors	48
5.1	Resolution versus Field Area	77
5.2	Schematic of Step-and-Repeat Aligner	79
5.3	Schematic of Electron Beam System	96
5.4	Various X-ray Lithography Techniques	102
5.5	Schematic of X-ray Lithography Using Synchrotron Radiation	104

ILLUSTRATIONS

5.6	Schematic of X-ray Stepper	112
5.7	Schematic of X-ray Reduction System	114
5.8	Schematic of Focused Ion Beam System	116
5.9	Schematic of Masked Ion Beam System	118
5.10	Schematic of Holographic Lithography System	123
6.1	Light Transmittance of Glasses	128
6.2	Photomask Fabrication Flow	130
6.3	Optical Photomask Fabrication Flow	130
6.4	E-beam Photomask Fabrication Flow	131
6.5	Phase Shifting Masks	133
6.6	Schematic of a Laser Pattern Generator	137
6.7	Die-to-die (Top) and Die-to-database (Bottom) Inspection	139
6.8	Schematic of a Focused Ion Beam System	141
6.9	Illustration of Clear and Opaque Mask Repair	143
7.1	Schematic of Typical Single Wafer RIE System	148
7.2	Schematic of Precision 5000 Etch Multichamber System	150
7.3	Schematic of Dry Processing Systems	152
7.4	Schematic of Beam-source ECR	155
7.5	Schematic of Multipolar ECR	156
7.6	Various Enhanced Designs (a) Triode, (b) HiRRIE, (c) MERIE	159
7.7	Schematic of Helicon Whistler Source	160
7.8	Schematic of Helical Resonator Source	160
7.9	Silicon Trench Structure	163
8.1	Schematic of Electron Beam Evaporation Method	175
8.2	Schematic of Ionized Cluster Beam Deposition Technique	176
8.3	Diagram of Planar Magnetron Sputtering Source	177
8.4	Schematic of Ion Beam Deposition Technique	177
8.5	Schematic of Advanced APCVD	179
8.6	Schematic of Hot-wall LPCVD Reactors	180
8.7	Schematic of Cold-wall LPCVD Reactors	182
8.8	Schematic of PECVD Techniques	183
8.9	Schematic of SACVD Techniques	184
9.1	Schematic of Excimer Laser	205
9.2	Comparison of Laser Photolysis and Pyrolysis	207
9.3	Schematic of Confocal Scanning Laser Microscope	209
9.4	Laser and White Light-based Particle Measurement Systems	213
9.5	Methods of Endpoint Detection	215

ILLUSTRATIONS xiii

9.6	Schematic of Direct Write Laser Pantography	219
9.7	Schematic of Laser Projection Microchemistry	220
9.8	Schematic of Intelligent Laser Soldering Process	222

TABLES

1.1	Levels of Integration of Dynamic RAMs	3
2.1	Comparison of DRAM Production Factors	22
3.1	Common Wafer Processing Chemicals	26
3.2	Photoresist Stripping Solutions	27
3.3	Wet Stripping Systems	29
3.4	Advantages and Disadvantages of Various Cleaning Methods	36
3.5	Chemical and Material Compatibility	38
3.6	Common Gallon Containers for Standard Chemicals	41
3.7	Common 55-Gallon Drum Liners for Semiconductor Processing Chemicals	42
3.8	Cost Analysis/Payback of Reprocessor	50
4.1	Gas Control System Issues	62
4.2	Purity Specifications of Specialty Gases	66
4.3	Range of Purity of CVD Gases	67
4.4	Potential Hazards of Processing Gases	71
5.1	Photolithographic Cost Analysis	80
5.2	Positive Photoresist Enhancement Methods	92
5.3	Commercial Photoresist Characteristics	94
5.4	Characteristics of X-ray Systems	101
5.5	Worldwide Synchrotron-based XLR	110
5.6	Conventional Resists Compatible with Soft X-rays	115
5.7	Comparison of Lithographic Techniques	122
7.1	Silicon Wafer Size Usage in 1988 and 1991	146
7.2	Characteristics of Dry Processing Systems	153
7.3	Comparison of RIE and ECR	157
8.1	Deposition Techniques for Silicide Formation	189
8.2	Comparison of Film Properties for PECVD Passivation Materials	195
9.1	Lasers in Semiconductor Processing	205

PREFACE

This book examines and projects the technologies involved in the fabrication of very large scale integration (VLSI) semiconductor devices. Specifically, it deals with their likely developments, why and when their introduction will occur, what problems and choices are facing users, and where the opportunities and pitfalls lie. It is written from an industry perspective.

The more than 200 processing steps required to make an integrated circuit (IC) entail the use of chemicals and equipment, all housed in a contamination-free environment, or cleanroom. In this book, the most significant technological trends in products, processing, and applications are explained, including a description of the suppliers of chemicals (liquids, gases, sputtering materials) and equipment (lithography, mask making, plasma etching, and thin film deposition).

The focus is on the most rapidly changing advances in the semiconductor industry, not necessarily on the most important processing steps in IC fabrication. For example, issues such as GaAs wafer growth largely deal with improved purity and larger growth of wafers to 4″ and greater, whereas the mechanisms of Czochralski (CZ) growth for ICs and Bridgman (horizontal) for LEDs is firmly established. The same is true for Si, where discussions on transverse and axial magnetic fields in CZ growth have been resolved, and growth of 8″ (and larger) wafers is the main issue. Similarly, the choices are between vertical and horizontal diffusion furnaces, different types of ion implantation equipment, and furnaces and rapid thermal processing (RTP) for annealing.

The technologies discussed in this book are rapidly changing, often with several options currently competing to do the same job. In the lithography chapter, for instance, optical methods compete with non-optical methods. Within the optical methods, scanners compete with step-and-scan, which in turn competes with steppers. Within steppers, G-line competes with I-line, which competes with deep UV. In deposition, CVD competes with PVD. Within CVD, LPCVD competes with APCVD, which competes with PECVD. The same is true in the mask making chapter, with E-beams competing with lasers for mask making, and lasers competing with ion beams for mask repair. In the materials chapters, one-gallon containers are competing with 55-gallon containers, which are competing with truck loads that are distributed via piping in the cleanroom. The way these materials are purified and particulates removed represents one of the biggest challenges to high yields in the fab. These yields are also affected by gas purity and are also a function of cleanroom type.

The main purpose of this volume is to assist the reader in evaluating the spectrum of products, packaging, and dispensing systems available for use. Criteria for selecting vendors as well as chemical delivery and dispensing systems that will meet specific requirements are also outlined. Suppliers can also gain insights into future user needs for purposes of long-range planning, new product development, and product improvement.

In sum, this extensively illustrated book addresses the strategic issues and major product trends in the IC industry. As such, it should be of particular interest to executive personnel of semiconductor manufacturing facilities, strategic planners of semiconductor facilities, buyers of chemicals and equipment for the semiconductor industry, product planners of chemicals and equipment to the semiconductor industry, chemical and equipment suppliers to the semiconductor industry, investment analysts, and students.

ACKNOWLEDGEMENTS

The following figures and tables have been reprinted in part or in their entirety by permission of the publisher and/or the copyright holder.

Figures 1.1, 5.2, 7.3, 8.4, 9.5 and Tables 1.1, 3.1, 3.6, 3.7, 4.1, 4.4, 5.2, 5.3, 5.4, 7.1, 7.2, 9.1 from The Information Network, San Francisco, California

Figure 2.1 from *CleanRooms*, March 1992, published by Witter Publishing Company, Inc. Permission also granted by Bechtel Microelectronics, Beaverton, Oregon

Figures 2.2, 2.3 and Tables 3.2, 3.3 from *Semiconductor International*, 1986, published by Cahners Publishing Company

Figure 2.4 from Asyst Technologies, Inc., Milpitas, California

Figures 2.5, 2.6, 3.2, 3.3 from Hewlett-Packard Company, Santa Clara, California

Figure 2.7 from Shimizu Corporation, Tokyo, Japan

Figures 2.8, 5.5, 5.6, 5.9, 5.10, 7.4–7.8 and Table 5.7 from *Solid State Technology*, 1986, 1990, 1991, published by PennWell Publishing Company

Figure 3.1 from *Microcontamination*, 1983, published by Canon Communications, Inc.

Figure 3.4 and Table 3.8 from Athens Corporation, Oceanside, California

Figure 5.1 from SVG Lithography, San Jose, California

Figure 5.3 from *Journal of Vacuum Science and Technology*, 1973, published by the American Institute of Physics

ACKNOWLEDGEMENTS

Figure 5.4 from *Lasers and Optronics*, 1987, published by Gordon Publications

Figure 5.7 from American Telephone and Telegraph Company, 1990, Holmdel, New Jersey

Figure 5.8 from FEI Company, Beaverton, Oregon

Figures 6.1–6.4, 6.7 from Tektronix, Beaverton, Oregon

Figure 6.5 from Matsushita Electronics Corporation, Osaka, Japan

Figure 6.6 from Ateq Corporation, Beaverton, Oregon

Figures 6.8, 6.9 from Micrion Corporation, Peabody, Massachusetts

Figure 7.1 from Plasma-Therm, Voorhees, New Jersey

Figures 7.2, 8.8, 8.9 from Applied Materials, Santa Clara, California

Figure 7.9 from HoneyWell/GCA, Plymouth, Minnesota

Figure 8.1 from SRI, Menlo Park, California

Figure 8.2 and Table 2.1 from Mitsubishi Electric Corporation, Tokyo, Japan

Figure 8.3 from Kurt J. Lesker Company, Clairton, Pennsylvania

Figure 8.5 from Watkins-Johnson Corporation, Scotts Valley, California

Figure 8.6 and Table 8.1 from Varian Corporation, Palo Alto, California

Figure 8.7 from Genus Corporation, Mountain View, California

Figure 9.1 from Lamda Physik, Acton, Massachusetts

Figure 9.2 from Gould Electronics, Eastlake, Ohio

Figure 9.3 from Nikon Instrument Group, Melville, New York

Figure 9.4 from Climet Corporation, Redlands, California

Figure 9.6 from *Applied Physics Letters*, 1983, published by the American Institute of Physics. Permission also granted by Lawrence Livermore Laboratories, Livermore, California

ACKNOWLEDGEMENTS

Figure 9.7 from Massachusetts Institute of Technology, Lincoln Laboratory, Lexington

Figure 9.8 from Vanzetti Systems, Inc., Sun Valley, California

Table 3.4 from Research Triangle Institute, Research Triangle Park, North Carolina

Table 3.5 from Compass Publications, La Mesa, California

Table 4.2 from Matheson Gas, Secaucus, New Jersey

Table 4.3 from Solkatronic Chemicals, Fairfield, New Jersey

Table 5.1 from Ultratech-Stepper, Santa Clara, California

Table 5.5 from F. Cerrina, CX_RL, University of Wisconsin, Stoughton

Table 5.6 from Hampshire Instruments, Inc., Marlborough, Massachusetts

Table 7.3 from Electrotech, Bristol, UK

Table 8.2 from Novellus, San Jose, California

Chapter 1

IC INDUSTRY TRENDS

1.1. TRENDS IN IC PROCESSING TECHNOLOGY

From an economic standpoint, all ICs have a common need for arbitrarily small size and thus low cost per function. This demand for low cost and high performance requires:

- Smaller dimensions of details leading to much larger chip dimensions
- Higher yield of individual components
- Faster access time and reduced power

One characteristic of integrated circuit development is the recognition that a decline in fabrication cost can be achieved with an increase in chip complexity. In most VLSI applications in which circuits of high complexity are used, memories and microprocessors dominate and are generally treated as the leading indicators of IC advances. The devices that are currently being manufactured include:

- 16M Dynamic RAM (MOS)
- 1M Static RAM (MOS)
- 4M ROM (MOS)

1

IC INDUSTRY TRENDS

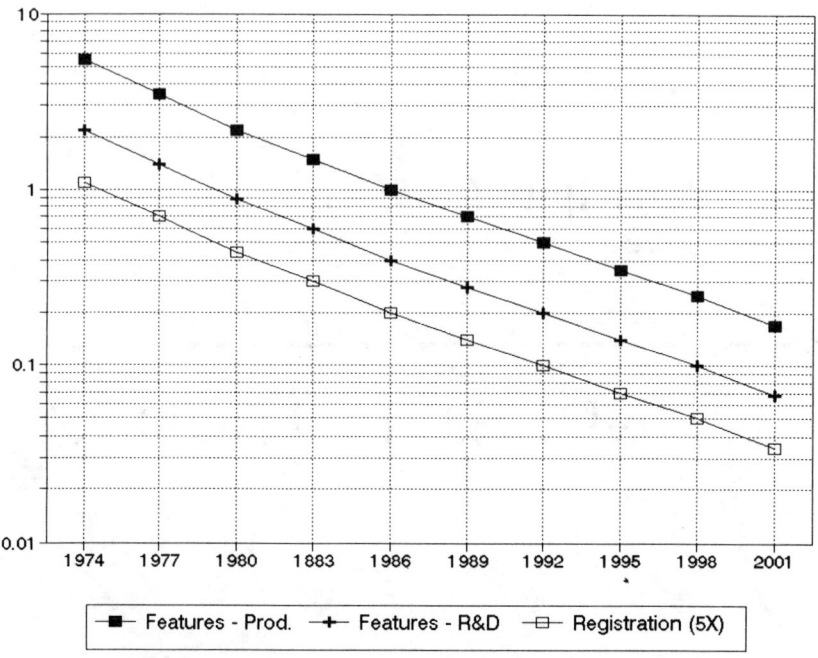

Figure 1.1. Minimum feature size for dynamic RAMs with time.

- 4M Flash EEPROM (MOS)
- 4M EPROM (MOS)
- 4M EEPROM (MOS)
- 32-bit Microprocessor

The reduction in the cost per function as a result of increased densities is exemplified in the dynamic RAM chip. Figure 1.1 shows the decrease in minimum feature size with time for R&D and production starting with Intel's 1103 1Kbit DRAM introduced two decades ago.

1 Mbit DRAMs have recently entered the commodity status and are in production by vendors at 20 million devices a month. 4 Mbit devices are in the production stage and Toshiba, with its trench cell design, has the lead. Other companies are following closely—Hitachi with its conservative stacked-cell design, and NEC with significant expertise in 1Mbit DRAM production. The 16Mbit device is in production at IBM.

IC INDUSTRY TRENDS

Table 1.1. Levels of integration of dynamic RAMs.

Year	Device	Linewidth (in microns)
1974	4 Kilobits	5.5
1977	16 Kilobits	3.5
1980	64 Kilobits	2.2
1983	256 Kilobits	1.5
1986	1 Megabit	1.0
1989	4 Megabits	0.7
1992	16 Megabits	0.5
1995	64 Megabits	0.35
1998	256 Megabits	0.25
2001	1 Gigabit	0.17

IC device density, based on the DRAM, has been doubling every 18 months. In the past two decades, technology has evolved from 4K DRAMs with 8 micron lines and spaces to 256K DRAMs with 1.5 micron lines and spaces as shown in Table 1.1. As this trend continues, 64M DRAMs with 0.35 micron minimum features will be commercially available in the next few years.

1.2. TRENDS IN CONTAMINATION

The yield and reliability of semiconductor devices is a function of particulate contamination in all stages of fabrication. Random defects on photomasks are the major source of yield loss in semiconductor devices, with particulates accounting for nearly half of these defects.

This book addresses the issues of contamination of Si and GaAs wafers from liquid and gaseous chemicals, de-ionized water, and ambient air. The latest developments in the identification, monitoring, and removal of these contaminants are described, together with the trends in usage.

Issues important to both user and supplier are described, focusing on a need to develop a synergy between these two groups. Most importantly, issues impacting users and suppliers in a competitive atmosphere with the Japanese are elucidated.

This book addresses cleanroom construction, filters, and wet processing systems. Also discussed are factors impacting these markets, such as the reduced usage of liquid etchant filters and wet processing systems because of the increased use of plasma etching.

1.3. TRENDS IN CHEMICAL USAGE

Chemicals and materials are used in every processing step in the fabrication of silicon and gallium arsenide integrated circuits. Technological advances in Si and GaAs ICs have resulted in more stringent requirements in the purity and quality of processing chemicals and materials for cleaning, etching, and deposition. As linewidths decrease, the level and size of contaminants in both chemicals and the manufacturing cleanroom become increasingly important as it directly impacts device yield.

The worldwide chemical and material market is also affected by these technological trends as new processes gradually replace old. Plasma etching (PE) and chemical vapor deposition (CVD) are two examples. Thus, the specialty gas segment will grow at the expense of liquid etchants and sputtering targets.

Because of competitive pressures in the IC industry, suppliers are often unaware of how their product is utilized in IC processing. As a result, they are unable to respond quickly to emerging technologies and products. This book attempts to bridge the gap between user requirements and supplier attitudes. This synergy is considered essential in order for both user and supplier to remain competitive in this fast-moving industry.

During the last 30 years, the semiconductor manufacturing industry has progressed from the production of relatively simple devices such as transistors and other discretes to the highly sophisticated and extremely complex very large scale integrated circuits that are being produced today.

Semiconductor fabrication techniques have changed consistently during this period. Almost every area in the manufacturing process has changed. Wafers are larger with higher and more uniform quality. Photoresists have been improved and larger selections are offered. Almost every piece of equipment has become more sophisticated from imaging equipment to furnaces to ion implanters. Automation is fast making inroads and already there is talk about "lights out" automation.

In recent years plasma etching and stripping has come into accepted usage and will continue to make in-roads as line widths decrease in size.

However, during the entire thirty years, chemicals have played a major role in semiconductor processing and that important role continues today. The chemicals offered today are certainly more chemically pure and contain less particulate contamination. The chemical companies are to be commended for offering much higher quality products at prices that are acceptable to the industry.

Although this improved quality is absolutely required by today's semiconductor technology, the basic concepts of chemical production, packaging, distribution and dispensing have not changed to the same degree that other

IC INDUSTRY TRENDS

parts of the manufacturing process have changed. However, during the last three years the most significant progress has been made. Improvements have been made in the basic manufacturing process, storage and shipping, filtration, availability of improved packaging, analytical capabilities and dispensing systems.

1.4. EQUIPMENT TRENDS

Each new generation of IC devices brings about a corresponding decrease in linewidths and minimum feature sizes. The technological trends and innovations in IC fabrication processes directly influences the market for processing equipment.

The primary objective of this book is to review the current issues dealing with processing equipment as applied to the manufacture of VLSI devices. The growth in the following segments will be detailed: (1) Lithography. (2) Mask making, Inspection, and Repair. (3)Plasma Etching. (4) Thin Film Deposition.

(1) The lithography market is the most competitive of all front-end semiconductor equipment markets, due to the high price of the equipment and the potential for high profit. The technological and market growth are established for proximity/contact aligners, scanning projection aligners, step-and-repeat aligners, E-beam systems, X-ray systems, and focused ion beam systems.

- Optical lithography will continue to dominate the lithography market.

- X-ray lithography continues to be pushed back by advances in optical.

- E-beam direct write will be relegated to a few prototype devices in a mix-and-match approach with optical.

- G-line steppers will reach their limit at 4 Mbit DRAMs, and will be replaced by I-line steppers for 4 Mbit and 16 Mbit DRAMs, as well as non-critical layers on 64 Mbit DRAMs.

- Excimer laser-based steppers will replace I-line steppers for 64 Mbit (0.35 µm resolution) and 256 Mbit (0.25 micron resolution) DRAMs beyond 1997.

- Japanese companies continue to dominate the stepper market, while U.S. companies face reduced market shares and liquidation.

(2) Mask-making needs in a VLSI facility are complicated by the high cost of capital equipment, estimated at more than $10 million. The need for

masks with smaller feature sizes and tighter specifications has required a high level of capital equipment purchases by the mask-making facility.

- Phase-shift masks and top-surface, multi-layer resists may extend I-line steppers even further than 0.25 µm resolution.

- Japanese companies continue to dominate the market, while U.S. companies face reduced market shares and liquidation.

- New systems will be developed for inspection and repair of X-ray and phase-shift masks.

(3) Plasma etching, which is rapidly replacing wet etching for the patterning of VLSI circuits, can be considered a mature technology. Nearly 40% of the thin films used in all types of integrated circuits are patterned by this method. Nevertheless, the technology is dynamic and new issues are brought to the forefront with each new generation of devices. As linewidths decrease further, new processes and equipment designs will be utilized. The increased usage of 8-inch wafers will further impact the worldwide equipment market. This market, dominated by batch systems, primarily with the hexode reactor design, has been redirected toward single wafer designs.

- Single wafer reactors will continue to grow at the expense of batch reactors.

- Single wafer, multichamber system will grow at the expense of multi-process chambers.

- ECR and MERIE (magnetically enhanced reactive ion etch) will gain momentum, as the lower operating pressures enable the plasma etch process to be extended to lower geometries.

- Low temperature, low ion damage, high selectivity equipment will be developed.

- Plasma stripping to be performed by single-wafer, downstream RF/microwave modules.

(4) Thin film deposition technology is in a state of evolution within the IC industry as a result of developments based on new interconnection materials. This market will see an evolutionary movement from evaporation to sputtering to chemical vapor deposition as the IC market moves from polysilicon toward refractory metal-based materials with submicron dimensions.

- CVD will be a driving force in cluster tools.

IC INDUSTRY TRENDS

- ECR CVD will grow, largely for its ability to deposit low-temperature, planarized dielectrics.
- Thermal TEOS/ozone BPSG processes continue to emerge.
- Tungsten CVD market will continue to gain entrants, as the market place heats up.
- CVD and PVD modules will be combined in cluster tools.
- PVD processes for multi-level metallization will drive the PVD equipment market.
- PVD will continue to be used for TiN, Ti-W, Mo gates, and silicides.

Chapter 2

CLEANROOMS

2.1. TYPES OF CLEANROOMS

Airborne particulates in the manufacturing environment must be reduced or eliminated to assure high quality and reliability of VLSI devices. These contaminants can be classified as (Hayakawa 1984):

- Gases—Nitrites, sulfites, carbon monoxide, and toxic gases
- Particulates—Dust, mist, smoke, and fumes
- Suspended Microorganisms—Algae, bacteria, protozoa, and viruses

Cleanrooms in the semiconductor manufacturing environment minimize particulate levels arising from the atmosphere, operating personnel, operating equipment, and tools. Air that passes through high-efficiency, 0.05 µm HEPA filters is virtually free of particulates (Ohmi et al. 1990), and the air downstream of the filter contains such a low particulate level that they do not register on a light-scattering particle counter. However, dynamic measurements during processing have identified as sources of contaminants:

- Skin scale
- Dandruff

- Bodily residues
- Garment fibers
- Silicon powder
- Residual resist fragments

The human body sheds two-tenths of an ounce of skin every day so that smocks and booties are not sufficient in Class 10 and 1 cleanrooms. Particles accounted for as much as 60% of the defects contributing to multiple failures during testing of 64K RAM wafers.

Class designations are a definition of the cleanliness of a cleanroom. However, below Class 100, officially defined as 100 particles of 0.5 μm or larger per cubic foot of purified air, the designation is meaningless. Class 10 cleanrooms definitions have merely been extrapolations of Class 100. One definition is 10 particles of 0.5 μm or larger per cubic foot. Another definition is 10 particles of 0.12 μm or larger per cubic foot. This discrepancy may be eliminated if the proposed definition by the RP-50 Committee of the Institute of Environmental Sciences (Mt. Prospect, IL) is accepted. Class 10 will be defined as:

- 10 particles 0.5 μm per cubic foot
- 30 particles 0.3 μm per cubic foot
- 350 particles 0.1 μm per cubic foot

2.1.1. Advantages and Disadvantages of Cleanroom Types

Cleanroom design in terms of laminar and nonlaminar air flow must be defined by the requirements of the device processing technology and include the degree of purity, operational control, and installation costs. There are two major types in the concept of the extensive cleanroom (Hayakawa 1984):

- *Horizontal Laminar Flow System.* Provides a predictable pattern of air flow. Air is brought into the cleanroom through a bank of HEPA filters occupying one wall and exhausted at the opposite wall. A Class 100 cleanliness level is usually obtained at the first workstation downstream from the filter, which degrades further downstream. The rate of recirculation of air is typically 200–300 ft^3 per hour at a discharge velocity of 90 ft per minute. The presence of furniture, equipment, and occupants will disturb the flow pattern, but the turbulences can be predicted and workplaces located is areas of the cleanroom so that contaminants generated will not affect other workstations.

- *Vertical Laminar Flow System.* Provides purified air from HEPA filters installed over virtually the entire surface of the ceiling through the lattice structure of the floor. Because of ceiling mounted lighting fixtures and other equipment, the flow is not perfectly laminar. However, a Class 100 environment can be maintained with flows as low as 50 ft per minute. The system has advantages over horizontal laminar flow systems, but it is more costly.

Advantages of extensive cleanrooms are:

- Simple design using standardized filter ceilings that permit the simple combinations with wall systems of the same modular dimensions
- Simple air-handling systems with constant temperature that does not require subdivision into control zones
- High degree of flexibility for the location of equipment and having intermediate partition walls
- Flexible layout of the system to meet the changing needs within the cleanroom

Disadvantages of this concept include:

- High flow rates coupled with high energy costs
- Extensive subterranean space under floors necessary for the return air plenum chamber
- Air cleanliness requirements cannot be differentiated within the cleanroom

With the clean tunnel concept, process equipment is arranged in double lines adjacent to the lateral walls, with a service and maintenance area on the other side of the wall as shown in Figure 2.1. The laminar flow is brought down as close as possible to the process equipment.

In the clean tunnel or bay system, laminar flow is maintained within the tunnel and can be started, stopped, and regulated individually within each tunnel. Maximum cleanliness is obtained directly under the tunnel, and each process can be isolated in its own chamber. Dollies or robots are used to move wafers between processes. The clean tunnel is not a fixed installation for obtaining a local clean environment, but a comprehensive technology adaptable to a variety of processes.

The system has a number of features described below (Hashimoto 1983):

- Initial costs are reduced 50%, and running costs to 45% compared to standard laminar flow cleanrooms.

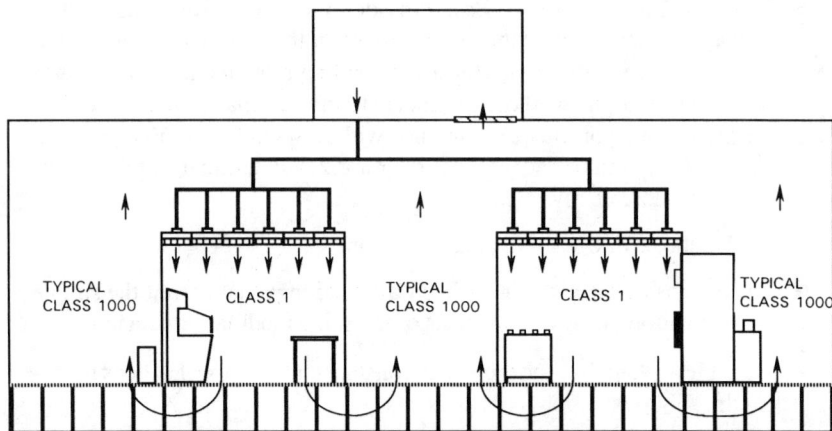

Figure 2.1. Cleanroom tunnel concept.

- Installation of clean modules measuring 40 feet long and 15 feet wide can accomplished in a little as 10 days without implementing rigorous contamination controls in the work area.
- Expansion can be implemented without shutting down the clean room.
- Cleanliness of each clean tunnel can be varied and reduced if necessary to a minimum, with the air supplied to the next module for re-use any number of times.
- Temperature and humidity within each module can be controlled individually depending on the load.
- Layout changes can be implemented easily.

Disadvantages of the tunnel concept are:

- Reduced flexibility to changing operational requirements and layout alterations
- Detailed planning is necessary for defining the layout of the facility, location of equipment, and material flow.

As a rule of thumb, cleanrooms <2,000 sq ft are more cost effective if modular in design because of the low engineering costs associated with this approach, cleanrooms >3,000 sq ft are more cost effective if "stick built", while those between 2,000 and 3,000 sq ft are approximately the same price to build.

The newest cleanroom configurations utilize full HEPA filter ceilings, an airflow of 100 fpm, tunnels measuring 12–14 feet wide and 8 feet long, and process equipment located either along the wall or accessed through the wall. Raised, perforated floors provide an air return that minimizes turbulence and horizontal cross flow.

2.2. AIRBORNE PARTICLE MONITORING

Monitoring of air purity, airflow velocity, air pressure, harmful gases, microorganisms, humidity, and temperature in the cleanroom is required in order to determine the cause of wafer defects during standard processing conditions. Continuous, real-time monitoring at each processing step enables accurate determination of the cause of wafer damage when it occurs. This type of monitoring equipment would include:

- Instantaneous particle counts of air adjacent to wafer
- Compact air sample intakes and detection mechanism so as not to interfere with processing
- Simultaneous recording of times and particle counts
- High particle detection accuracy

As discussed in Chapter 9, the optical particle counter is the most widely used method in cleanrooms. A light source illuminates a small viewing volume and a photodetector measures the scattered light from the individual particles as they pass through this illuminated volume. Lightscattering systems include monochromatic laser and polychromatic white light monitors. Both systems operate nearly identically; the difference between the two is that the laser system is able to concentrate more energy in a smaller spot.

The continuous-flow condensation nucleus counter operates by passing the gas through a heated alcohol vapor and a refrigerated condensor. Condensed alcohol droplets emerge and are counted by an optical system similar to that used in the method above. These droplets are able to condense on particles as small as 0.1 µm.

A large number of improvements have taken place with both laser and white light systems over the past few years, lowering the limits of sensitivity, and increasing repeatability. In principle, sensitivity limits vary from system to system, with the counting efficiency decreasing gradually rather than exhibiting a sharp step.

Difficulties arise when flow rates are reduced in order to increase sensitivity causing:

- High signal-to-noise ratio, equivalent to 1 to 10 particles per cubic foot, rendering counts in a Class 10 Cleanroom meaningless. New advances in electronics has resulted in counters with reduced signal-to-noise lev-

14   CLEANROOMS

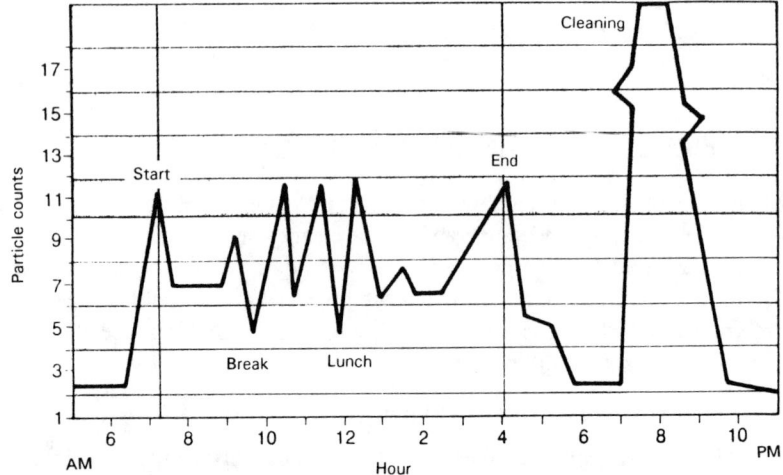

Figure 2.2. Contamination levels during cleanroom work cycles.

els, so that the counting efficiency may be even higher at lower flow rates. Existing counters in numerous cleanrooms, however, do not have sophisticated electronics, so that the counting efficiency is affected by the age of the of the counter, and whether any dirt has entered the system.

- At low flows, sampling must be made over longer time periods, up to 1.5 hours. This represents a problem with personnel activity, which is a major source of contamination within the cleanroom. Measurements, illustrated in Figure 2.2, show that contamination levels are directly related to personnel activity.

Condensation nucleus counters do not have size-discriminating capabilities and thus count all particles larger than 0.01 µm. This is a limitation that prevents its exclusive use in the cleanroom.

2.3. ELECTROSTATIC DISCHARGE

Mitsubishi has reported that 46% of the particles detected that were greater than 1 µm were of human origin. Particles this size are too large to pass through the HEPA filters and come from outside the laminar flow. Since particles cannot flow upstream against the current in a laminar flow, they are carried downstream on a laminar flow of clean air and exhausted. However, induction and static electricity force particles upstream in laminar flow clean

CLEANROOMS

rooms. Although the airflow may be superclean, use of high-efficiency filters alone does not satisfy the requirements of cleanroom technology (Inaba and Ohmi 1989).

Induction arises because air flow is not strictly laminar but subject to turbulence. This can be controlled somewhat by regulating the airflow configuration and altering the position of furniture, equipment, and areas of personnel workplaces. Static electricity can be generated by cleanroom garments made from poor conductors such as nylon and polyester. Surface charges can by monitored by a potentiometer and procedures should be implemented when readings exceed 5000V. Procedures identified by Motorola's Microprocessor Division to reduce static electricity include:

- Antistatic and carbon rails
- Antistatic floor treatment
- Antistatic spray on chairs, partitions, etc.
- Carbon-coated boxes
- Completely grounded automatic handlers
- Control of relative humidity to 40%
- Grounded work surfaces
- Ionizers
- Static-shedding bags
- Wrist straps

2.4. FUTURE TRENDS

A Class 1 cleanroom, defined as 1 particle 0.2 µm or larger per cubic foot of filtered air, can be achieved with a full HEPA filter ceiling and 480 changes of air per hour. This type of cleanroom, illustrated in Figure 2.3, consists of a 100% vertical laminar flows in a 12,000 sq ft cleanroom within a cleanroom with a total area of 25,000 sq ft. The other 12,000 sq ft consists of service aisles and a perimeter corridor.

2.4.1. Impact of Automation

In only two years, modular, multichamber, integrated process systems—cluster tools—have achieved worldwide recognition, and have become

16 CLEANROOMS

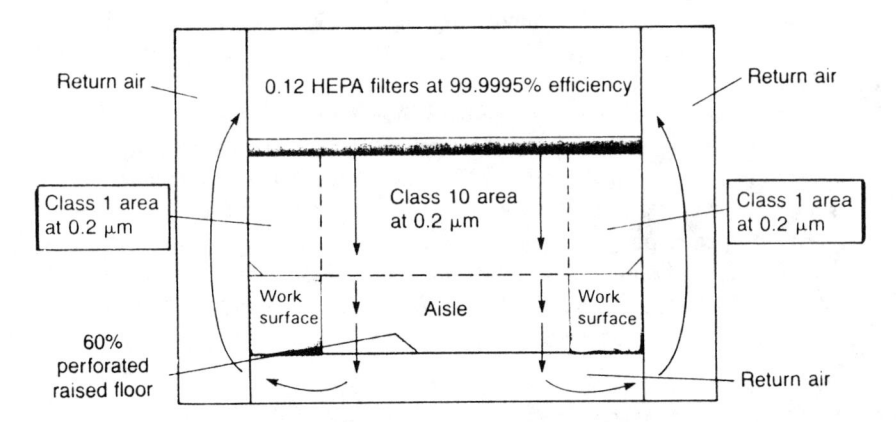

Figure 2.3. Class 1 cleanroom configuration.

widely accepted in the U.S. and Europe as a key concept for VLSI manufacturing.

The perceived advantages of cluster tools comes at a time when the semiconductor industry has been is a sustained recession, when rising costs has increased the cost of a 20,000 sq. ft. fab to $80 million, and semiconductor manufacturers are beginning to make purchases for the next generation of devices (64 Mbit DRAMs or equivalents), which will not be in full production until 1995.

The use of automation within a cleanroom can be integrated with protective transfer systems. The Standard Mechanical Interface (SMIF) (Parikh, 1984), shown in Figure 2.4, can achieve high purity levels in an impure environment created by cleanroom personnel. SMIF isolates the wafers from the fab operators and can be incorporated into existing as well as new fabs. Thus, the SMIF system can produce a Class 10 or better cleanliness in existing Class 100 or higher cleanrooms. An alternative to the large capital investment of the facilities and utility expenses associated with air-handling and temperature-control equipment in Class 1 and Class 10 cleanrooms, SMIF can achieve high purity levels in an impure environment created by cleanroom personnel.

Studies at Hewlett-Packard have shown that SMIF boxes successfully reduce particulate levels on wafers (Gunawardena et al. 1985). As shown in Figure 2.5, SMIF boxes reduced particulate levels from 60 particles per wafer pass (PWP) to 20 PWP for one piece of processing equipment located in an aisle. In Figure 2.6, particulate levels were reduced from 4 PWP to 2.2 PWP for a piece of equipment located inside an environmental chamber.

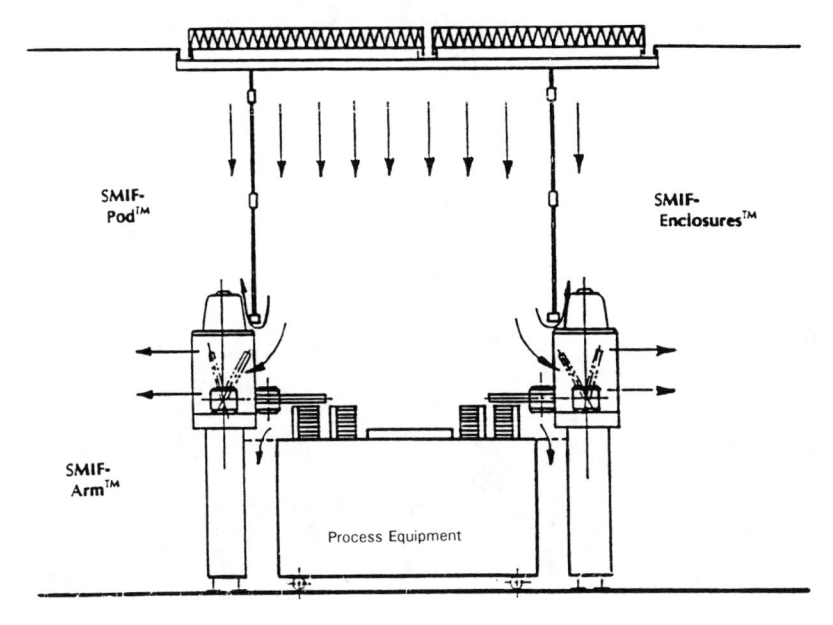

Figure 2.4. **SMIF concept.**

As shown in Figure 2.7, SMIF performance in a Class 20,000 is superior to conventional cassette handling in a Class 10 ambient.

In addition to SMIF, LETI (Grenoble, France) has introduced the concept of WAFEC (Wafer Air Flow Environment Container) (Douche, 1990) illustrated in Figure 2.8. LETI's concept improves localized environmental control by:

- Improving wafer protection within the process equipment by reducing the number of interfaces between the container and the equipment reactor.

- Introducing the notion of close confinement, which maintains the wafers in an ultrapure air flow at the input/output ports in regions between the ports and up to the reactors.

- The close confinement concept applied to the container, whereby the air flow guarantees tightness of the box while dynamically ensuring protection of the enclosure against contamination.

WAFEC implements close wafer protection by an ultraclean air flow that dynamically guarantees tightness of containers. Close confinement to

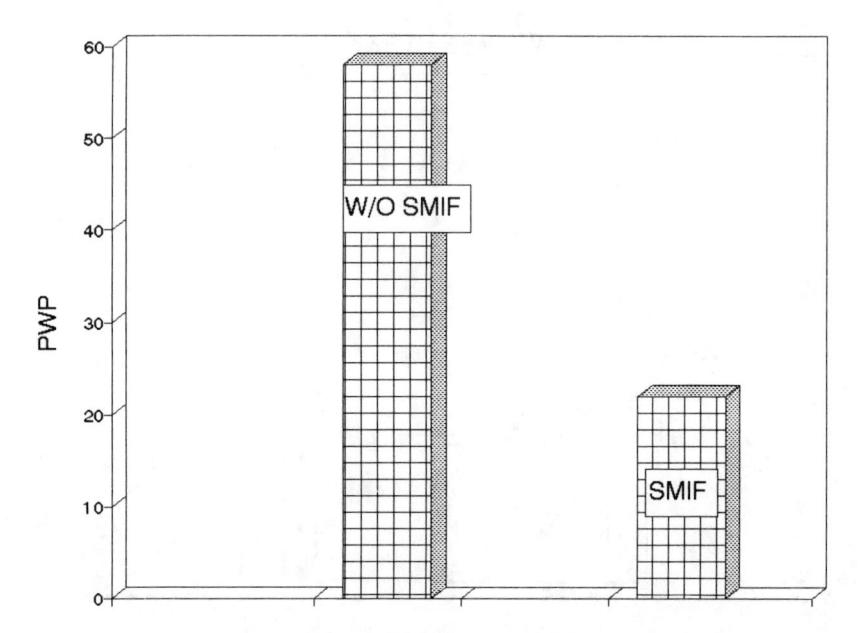

Figure 2.5. Contamination levels with and without SMIF box with equipment in aisles.

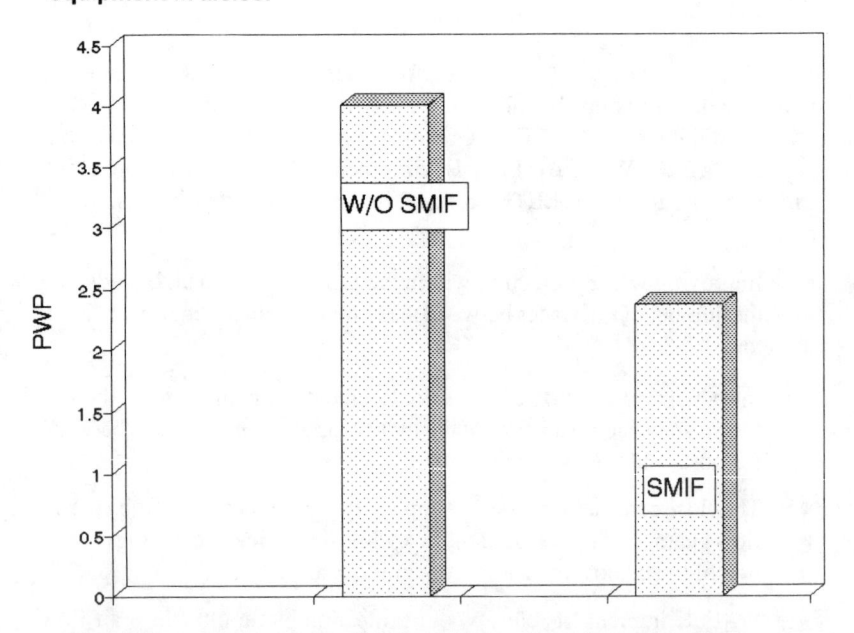

Figure 2.6. Contamination levels with and without SMIF box with equipment in environmental chamber.

CLEANROOMS

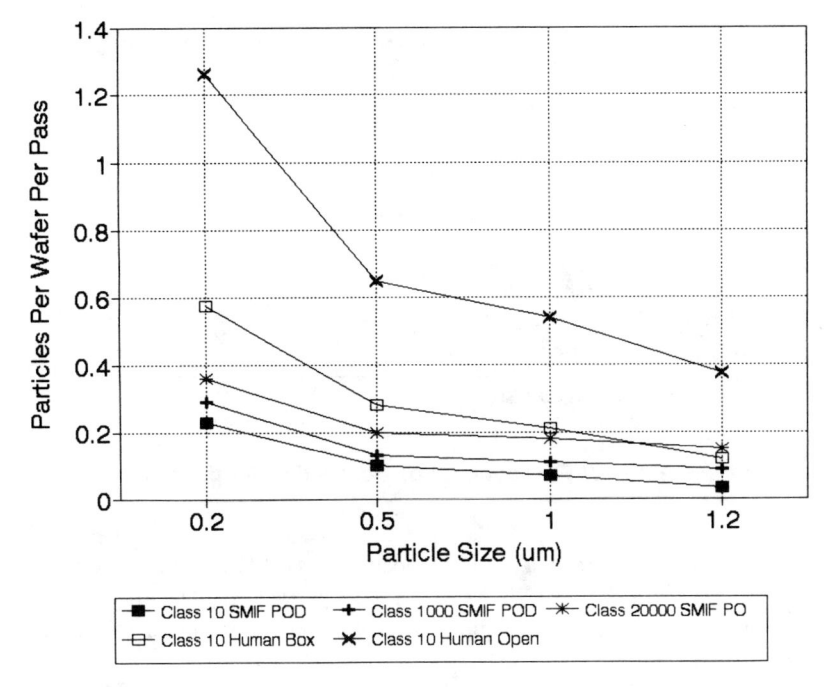

Figure 2.7. **Air cleanliness measurements with and without SMIF boxes.**

the equipment reaction chamber is ensured by blowing air through mini-ceilings laid out at carefully chosen points on the wafer path. The container is fitted with a mechanism that enables the cassette to be directly placed on or taken up from the equipment lift.

A specific base, accurately positioned on the equipment, acts as a mechanical interface between the container and the equipment. Prefiltered air is injected into the container through a connection secured to the base when the container is on either the equipment or storage cabinet.

Tests carried out at LETI indicate that within the container, the equivalent environment is better than Class 0.1.

In order for any advanced concept to be economical, there will be a trend towards smaller fabs when the impact of automation will be in full swing. The current status of automation is at the level of "islands of automation" with clusters of processing equipment such as a photoresist coat/bake system attached to a wafer stepper which is attached to an automatic feed inspection station. The problems of linking several pieces of equipment together are di-

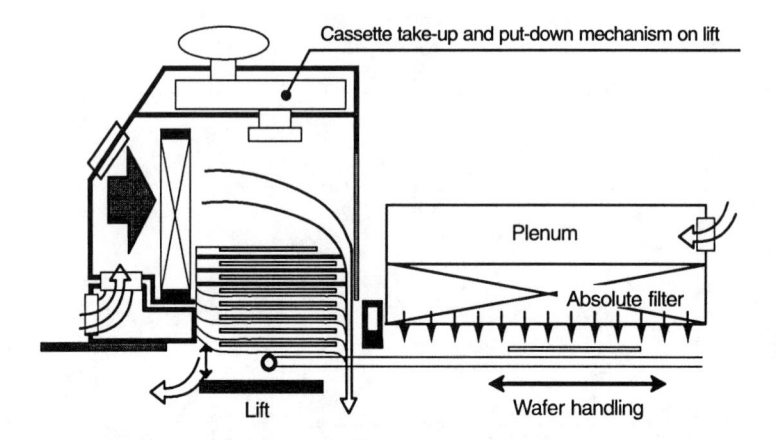

Figure 2.8. WAFEC concept of localized wafer protection.

rectly dependent on the mean time between assists (MTBA) of each system. Monster fabs with 10,000 to 20,000 wafer starts per week contain a dozen wafer steppers. Linking all this equipment would be an insurmountable challenge.

Along with the trend toward cleaner manufacturing environments comes a concomitant trend in advanced techniques for monitoring airborne contaminants. Federal Standard 209C requires a minimum of 400 sample locations to be tested in order to certify a 10,000 square foot cleanroom. Additionally, more than 12 monitoring locations may be required to properly monitor particulates on a daily basis.

Proper maintenance of Class 1 and Class 10 cleanrooms requires advanced monitoring techniques capable of not only particles as small as 0.1 µm, but relative humidity, temperature, airflow, static pressure drop, and traffic counts. System sophistication has progressed to a point where a closed-loop feedback control system is made to the air handling equipment.

Monitors based on laser light, with measurement capabilities of 0.1 µm at low flow rates will become the predominant technology, despite laser replacement costs of up to 15 times greater than white light systems.

High Yield Technology (Mountain View, CA) has developed a monitor that uses a net of laser light to monitor up to 100 passing particles/sec as small as 0.5 µm. The small size of the system—3 × 1.5 × 5/8 in.—enables it to be placed in small spaces and hostile environments generally occurring in process equipment. Particles passing through the laser net are measured; particles passing parallel to the net are not. Thus, the origin of the particulates can be determined by merely positioning the system.

Cleanroom personnel contribute approximately 30% of particulate contamination. Other sources of contamination arise from (Tolliver 1984, Gunawardena et al. 1985) (Ervin and Berger 1986):

- Inactive process—30%
- Equipment—25%
- Room environment—10%
- Wafer storage—5%

where inactive process contamination arises from waiting at the input/output ports of equipment and transit to the reactor.

In addition to the effort of removing personnel from the cleanroom in the much heralded "Factory of the Future," efforts to reduce these contamination levels are directed on three fronts: through the design of support areas adjacent to the cleanroom; through a complex regime of proper gowning of the worker; and in the design cleanroom garments.

The first two issues are interrelated and a synergistic effort must be established between the cleanroom supplier or designer and the semiconductor manufacturer. For example, if the removal of street clothes is mandated by the user, separate women's and men's locker rooms must be installed and maintained. The movement of workers must also be addressed. Traffic patterns must be in a one-way direction to avoid cross contamination.

The effectiveness of cleanroom garments, i.e., the ability to keep out operator-generated particulates, is both a function of the space between bundles of fibers (pore area) and the interstitial space between individual fibers comprising the yarn. As early as 1984, it was discovered that the ratio of free space to occluded space in fiber bundles determines the filtration rate, while the absolute value of the fiber diameter determines the particle removal efficiency. Since that time, efforts have progressed toward the development of new fibers with reduced particle penetration—achieved by modifying the yarn preparation, weaving, and finishing.

The overall efforts of more than 600 companies supplying materials and services to the cleanroom industry are providing efficient control of contamination levels that directly affects device yield. It is obvious that a user-supplier synergy must be developed and maintained for future advances in cleanroom technology in the path of ever decreasing device feature sizes.

2.4.2. Impact of ASIC Devices

The fast turnaround times and small production lots of application specific integrated circuits (ASICs) is further directing the size of cleanrooms. Mon-

Table 2.1. **Comparison of DRAM production factors.**

Factors	256K	1M	4M
Type	NMOS	CMOS	CMOS (trench or stack)
No. of masks	9–10	17–19	20–25
Total process steps	200	350	450
Test time (sec.)	60	120	240
Chip size (ratio)	1	2	3
Logical no. of chips/wafer (ratio)	3	2	1
Good dies/lot (ratio)	6	2	1
Cleanroom class	100	10	1
Cost ratio (on a mature production)	1	4	10

ster fabs costing up to $200 million are not economical for ASIC production. Neither are they sufficiently flexible for controlling small prototype lots or quick conversion to different designs.

The small size of these cleanrooms offers a number of advantages:

- A fab capable of a throughput of 5000 wafer starts per month can cost as little as $20 million, due to the reduced number of mechanical and electrical systems.
- ASIC fabs can be designed and built in as little as six months.
- The modular design of some of these fabs enables them to be constructed at the cleanroom manufacturer's site, and transported by truck. Costs are lower and existing fabs can be demolished while others are being erected.

2.4.3. Impact of VLSI Devices

As devices increase in complexity, the smaller feature sizes will dictate an increase in the usage of Class 1 cleanrooms to minimize the generation of killer defects. Shown in Table 2.1 is a comparison of DRAM production factors. While a 1 Mbit DRAM can be readily fabricated in a Class 10 cleanroom, increases process complexities of a 4 Mbit DRAM will necessitate the need for a Class 1 cleanroom.

2.5. JAPANESE CLEANROOMS

The Japanese goal of a virtually zero-contaminant environment utilizes cleanrooms as one of its most potent weapons to obtain high yields and low-cost devices (Edmark and Juackenbos 1984). Japanese cleanrooms are smaller than their U.S. counterparts, making it easier to control the particulate levels. All surfaces throughout the entire wafer fab building are constantly cleaned. Process equipment is thoroughly cleaned, tested, and re-cleaned with sticky tape before it is brought into the cleanroom. Japanese semiconductor manufacturers use solar cells in their equipment to remove particulates. A solar cell is placed within the process chamber of a system. When the cell is charged, it attracts airborne particulates that are then permanently removed from the chamber.

Japanese cleanroom workers take showers on their own time before getting into special throwaway undergarments and bunny suits. Workers leave street shoes at the entrance to the building and wear indoor shoes throughout.

The Japanese have recognized that particulate contamination is only part of the problem of cleanroom contamination and low wafer yields. The quality of the cleanroom air is monitored, and hydrocarbon concentrations are maintained at less than 0.1 ppm. Researchers in Japan have discovered that a significant factor in low yields is adsorption of molecules on the wafer from materials in the cleanroom—materials such as walls, windows, floor gratings, and filters.

These adsorbed materials, once on the wafer, are difficult to remove. Ultrapure DI water and gases could control the problem. Also, studies have shown that low-energy argon ion bombardment removes the adsorbed molecules and does not harm the wafer.

Thus, total cleanroom air quality has been an ongoing effort in Japan. In the U.S., this has not been recognized, and cleanroom air quality is measured by particulates, temperature, and humidity.

The Japanese have now begun to commercialize on their cleanroom techniques. Hitachi Plant Engineering and Construction Co. in 1989 began marketing a cleanroom suitable for 0.4-µm linewidth feature sizes. The prefabricated cleanroom measures $2 \times 2.4 \times 3$ meters plus an airlock room of $1 \times 1.5 \times 2$ meters.

Temperature inside the cleanroom is kept at a constant 23°C ±0.02°. Precise temperature control is vital at these feature sizes, as a slight change in temperature may cause expansion or contraction of critical materials or upset sensitive equipment.

The degree of cleanliness is 1 particle, 0.5 µm or smaller per cubic foot of space. Air pressure can be varied between 1,030 and 1,040 millibars, and within a ±0.3 millibar accuracy.

REFERENCES

Cheung, S.D. and R.P. Roberge, 1986: "Measurement of particles in IC process equipment," Proc. Microcontamination Conference, Santa Clara, CA, November 18–21, pp. 130–146.

Douche, C., 1990: "Wafer confinement for control of contamination in microelectronics," *Solid State Technology* **33** (8): S1–S5.

Edmark, K.W. and G. Quackenbos, 1984: "An American assessment of Japanese contamination-control technology," *Microcontamination* **2** (5): 47–53.

Ervin, R. and J. Berger, 1986: "Contamination in semiconductor processing," Proc. 32nd Annual Tech. Mtg., IES, pp. 424–426.

Gunawardena, S., U. Kaempf, B. Tullis, and J. Vietor, 1985: "SMIF and its impact on cleanroom automation," *Microcontamination* **3** (9): 54–62.

Hashimoto, T., 1983: "Challenging class 0 problems in superclean environments for 1M-bit devices," *JST News* **2** (2): 20–26.

Inaba, H. and T. Ohmi, 1989: "Influence of electrostatic charge," Proc. of 9th Symposium on Ultra Clean Technology, Tokyo.

Hayakawa, I., 1984: "Air purification for cleanrooms," *JST News* **3** (4): 15–20.

Ohmi. T., Y. Kasama, K. Sugiyama, Y. Mizuguchi, Y. Yagi, H. Inaba, and M. Kawakami, 1990: "Controlling wafer surface contamination in air conditioning, particle removal subsystems," *Microcontamination* **8** (2): 45.

Parikh, M. and U. Kaempf, 1984: "SMIF: A technology for wafer cassette transfer in VLSI manufacturing," *Solid State Technology* **27** (7): 111.

Tolliver, D.L., 1984: "Contamination control: New dimensions in VLSI manufacturing," *Solid State Technology* **27** (3): 129–137.

Chapter 3

LIQUID CHEMICALS

3.1. TECHNOLOGY ISSUES

3.1.1. Acids and Solvents

The increasing complexity of semiconductor devices dictates the use of dry processing methods for pattern delineation. Dry processing techniques have replaced wet etching and stripping of oxide and nitride films in most LSI devices. The use of dry processing replaced wet chemical etching for metallic gate and interconnect materials in most LSI and VLSI devices in 1986. Nevertheless, wet chemical etching is still used in more than 65% of IC devices and will continue to be required for mask and substrate cleaning.

Chemicals are used in a variety of processes in semiconductor manufacturing as listed in Table 3.1. All are used in IC manufacturing to varying degrees depending on the type of device architecture, cleaning-process chemistry and frequency of use, etching-process chemistry, method of thin film deposition, use of wet or dry etching, use of wet-etching machines (cassette-to-cassette, automatic processing) or wet etching systems (dip tanks), size of wafers, and number of wafer starts.

Table 3.1. **Common wafer processing chemicals.**

Sulfuric acid	Resist stripping—combined with peroxide in piranha bath
	Etching—combined with HF and nitric for Si
Hydrofluoric acid	Etching—combined with ammonium fluoride to form buffered oxide etchant (BOE) for oxide
	Etching—combined with nitric and acetic to form mixed acid etchants (MAE) for Si
Hydrochloric acid	Cleaning—key component in RCA clean for metal ions
Nitric acid	Etching—used in MAE and other etchants
	Stripping—in concentrated form for resists
Acetic acid	Etching—in MAE for etch rate control
Phosphoric acid	Etching—in phosphoric aluminum etchants
Ammonium hydroxide	Cleaning—key component in RCA clean
Hydrogen peroxide	Resist Stripping—combined with sulfuric
	Cleaning—component in alkaline and acid RCA
Ammonium fluoride	Etching—primary component in BOE for oxide
Solvents	Rinsing—general purpose cleaning and rinse

The most common cleaning solution is the RCA Clean; a solution of sulfuric acid, hydrofluoric acid, hydrochloric acid, ammonium hydroxide, and hydrogen peroxide (Kern and Puotinen 1970). This combination of chemicals serves several functions: a hydrogen peroxide/ammonium hydroxide solution reduces the hydrophobic nature of the wafer so that acidic solutions can penetrate; a hydrochloric acid/ hydrogen peroxide solution acts as a chelating agent to remove any inorganic residue left after acid cleaning.

Wet etching and photoresist stripping are operations that are undergoing the fastest conversion to dry processing. Dry etching has the advantage of high resolution patterning to submicron features. An additional advantage of dry processing is that most of the liquid chemicals are regarded as hazardous and require special handling and disposal procedures.

3.1.2. Resists

The choice of photoresist strippers depends on the photoresist and the materials on the surface of the wafer. The most commonly used examples are listed in Table 3.2 (Peters 1992). Sulfuric acid is the most widely used stripper with non-metals (oxide, nitride, poly) on the surface. Used at temperatures of 120° to 140° C, the reaction of the sulfuric acid with resist results in the formation of carbon. Oxidants such as ammonium persulfate (SA80) and hydrogen peroxide are added to the acid to convert this carbon to carbon dioxide. This is a universal stripper called Piranha etch (4:1 sulfuric:peroxide) and can be used

LIQUID CHEMICALS

Table 3.2. **Photoresist stripping solutions.**

Company	Model	Features
ACSI	ST–22	Non-corrosive to Al/Al alloys; strips RIE, uv cured resists
ACT	ACT–CMI	Non-NMP based; electrogalvanic corrosion inhibitor
BASF	NMP	Non-corrosive; 91° flash point
J.T. Baker	PRS–3000	Non-corrosive; 103° flash point; stripper return program
Cyantek	Nanostrip	Stabilized piranha with longer bath life and lower strip temperature
Dynachem	S–46	Non-NMP based solvent; non-corrosive; 85° flash point
EKC	Posistrip 830	Non-corrosive to Al/Al alloys; chemical recycling program
EMT	EMT 300	Non-NMP based; non-corrosive to Al/Al alloys
Hoechst	AZ 400T	Getters surface metal ion contamination; > 85° flash point
KTI	NovaStrip	Does not contain phenols, phosphates, chromates, or fluorides
OCG	Microstrip	Mixture of NMP and aminoethoxy ethanol
Shipley	Microposit	Low metal ion content; > 85° flash point
Silicon Valley Chemlab	SVC 150	Contains no phenols or halogenated hydrocarbons; 115° flash point
Tokyo Ohka	106	Contains no phenols or halogenated hydrocarbons; 89° flash point

with both positive and negative resists. However, the need to continually replenish the oxidant in the bath has encouraged users to purchase chemical reprocessors.

Moreover, organic-based solvents have been specifically designed to remove tough resists, such as heavily implanted resists and films following poly etch or aluminum etch. Many organic strippers are based on the industry standard solvent NMP (n-methyl-pyrrolidone), which are biodegradable and disposable by dilution in water. The operating temperature of the stripper is important: strippers with higher solvent strength are able to strip tough resists at lower temperatures. Solvent-based strippers are also inert to aluminum and its alloys.

Wet etching and photoresist stripping are operations that are undergoing the fastest conversion to dry processing. Dry processing has the advantage of high resolution patterning to submicron features. An additional advantage of dry processing is that most of the liquid chemicals are regarded as hazardous and require special handling and disposal procedures.

Dry-processing techniques for photoresist stripping have no distinct advantage over wet processing except for the handling and disposal problems. In fact, metallic ions present in photoresist are not removed by dry processing (Fujimura and Yano 1986). Wet stripping, more cost effective than dry, can remove these contaminants. As a result, plasma stripping of photoresist is used when the resist has been exposed to high energy ionic bombardment and proves too tenacious to be removed by wet chemical treatment.

Dry resist strippers were initially manufactured in a barrel configuration operating at 13.56 MHz. Because of the potential of high radiation damage, many of the newer dry strippers are specifically designed to handle tough resists. Single wafer systems are manufactured by Canon, Fusion Semiconductor Systems, GaSonics/IPA, Matrix Integrated Systems, Mattson Technology, Plasma System Corp., Samco, STS, Ulvac, and Yield Engineering Systems.

Dry stripping only partially removes residues, making latent particles harder to remove. Many companies use a multi-step process—dry stripping followed by a wet process. Aggressive wet stripping systems that employ megasonics or ultrasonics are often used to dislodge residues. These spray processes remove residues better than immersion methods, and at the same time reduce the likelihood of particle redeposition. Wet stripping systems are listed in Table 3.3.

Positive resists can achieve aspect ratios (rates of thickness to resolution) of 1:1; i.e., 1-μm thick layer capable of 1-μm resolution. Negative resists are capable of aspect ratios of only 1:2. For this reason, positive photoresists have made great inroads into the resist market (Jain 1990). With LSI and VLSI trends toward sub 0.5-μm resolution, positive resists have overtaken negative resists in terms of market share. Negative resists are still being used for low-density ICs and discrete devices and will continue to be used until these devices are no longer produced since conversion to positive resists would necessitate conversion of lithographic masks. Positive resist usage also predominates in metallization patterning because of the superior adhesion to these films.

Multilayer resist techniques can alleviate problems resulting from device topography, reflections from the substrate surface and standing wave effects (Ong and Hu 1984). In addition, novolac-based resists have a commercial limitation of 1 micron.

Anti-reflective coatings (ARC) also reduce internal reflections and standing waves in the resist and promote better adhesion. ARC is spun on the

Table 3.3. **Wet stripping systems.**

Company	Model	Stripping Method	Features
Athens	Piranha	I	Recycled sulfuring acid/oxidant
Bjorne Enterprises	Panda 9000	I,S	High-pressure, all-enclosed system for strip, rinse, dry
CFM Technologies	Full Flow	I, FF, M	Single chamber for strip, rinse, IPA dry; 300 wafers/hr
Convac APT	APTCON 3000	S	Combines coat, develop, etch, strip, and clean modules
FSI International	Mercury MP	S,M	Fully automated, multi-cassette system; 120–300 wafers/hr
Hamatech	HMR 90	S	Strips wafers and masks; 0.1 micron filtration
Integrated Circuit Dev.	**	I,U	Teflon and quartz baths; heaters
Lufran	Teflon Bath	I	High temperature filtered etch bath
J.M. Ney	ProSONIK	I,U	Ultrasonic frequency; 150 wafers/hr
Probe-Tronics	RE–200	I	Variable rotation and rapid quench; 200 wafers/hr
Pure Aire	**	I,U	Cascade or overflow rinsers; flexible modular system
Reynolds Tech	PWS–172–PS	I,M,U	Recirculation between tanks minimizes chemical usage
Santa Clara Plastics	8500	I,M,U	Fully automated dry-to-dry strip: SMIF compatible
Semiconductor Process Equip.	**	I,S,U,M	Automation, recirculation, filtration; flexible modular
Semitool	Spray Solvent	S	Dry-to-dry process; 75–125 wafers/hr
Solid State Equipment	175	S	Stripper/cleaner uses 0.1-micron filtered acid solution
Universal Plastics	UP–AWPS	I,M,U	Fully automated or manual in high-temperature baths
Verteq	**	I,M	Megasonic agitation for minimal damage
Wafab	Chemkleen	I,U	Recirculation bath with temperature

** = varies; I = immersion; S = spray; FF = fluid flow; M = megasonic; U = ultrasonic

substrate and imaged, developed, and removed with the upper layer photoresist. The only additional steps added to the photoresist process are coating and baking the ARC. ARC has been shown to eliminate 96% of all reflected light at 436 mm and provides the photolithography process with higher resolution and greater exposure latitude. Photoresist resolution with ARC is as good on 5X wafer steppers as photoresist alone on 10X steppers.

Contrast enhancement coatings have recently been developed by GE, making it possible to produce submicron circuitry with conventional techniques and equipment. The material is applied to the wafer at the beginning of the fabrication process. When a circuit image from an optical printer is focused on the coated wafer, a photo-bleachable dye in the coating creates a window, exposing a sharply defined area of the underlying photoresist. The coating creates its own mask on the wafer with more resolution than the projected mask. The processing requires two additional steps, one to apply the 0.3 µm-thick coating with conventional photoresist dispensing equipment and a second to remove the coating by an additional stripping solvent.

Polymethyl methacrylate or PMMA is the benchmark of electron beam positive resists because of its high resolution (0.05 µm), its process tolerance, and the minimal formation of swollen surface layers during dissolution. Negative features include low dry-etching tolerance and low sensitivity (80 microCoul/cm^2). Optical resists exposed with an electron beam can form either positive or negative images depending on the developer polarity. These materials can have a sensitivity of 30 microCoul/cm^2 and a resolution of 0.5 µm (Watts 1984).

Inorganic resists and onium-salt photosensitized systems are two alternative resist systems. The inorganic resist is a GeSe layer on which AgSe is evaporated. On irradiation, the Ag migrates into the GeSe, forming a layer that is developed to form a negative image. Although the resolution is high (<0.5 µm) the high sensitivity (500 microCoul/cm^2) is a limitation (Tai *et al.* 1982). The onium-salt system relies on a photochemical reaction of the onium salt into a strong acid that then initiates polymerization, degradation, or hydrolysis reactions (Ito and Wilson 1982).

3.2. PURITY REQUIREMENTS

3.2.1. Purification Methods

Suppliers of chemicals are focusing on reducing the impurity levels of their products. "Electronic" and "Semiconductor" grade acids and solvents are available with low concentrations of cations and anions that could degrade device performance. Transition metal impurities, for example, are fast diffusers and readily move into the silicon wafer lattice, replacing silicon ions at lattice sites and changing the bandgap of the silicon. These new sites act as a generation-recombination center and result in a decrease in the minority carrier lifetimes and leakage currents at p-n junctions (Duffalo and Monkowski 1984).

Trends for Purity—Trace Elements

In the first-half of the 1990s, the trace element standards will evolve to two standards projected below:

- *Super Premium.* Each individual trace element at less than 1 ppb with a total trace element of less than 10 ppb. Select elements such as Na, K, Li, Fe, Ca, Al, As, B, P, and Sb may have a maximum of less than 0.2 ppb.

- *Premium.* Each individual trace element at less than 5 ppb with a total of less than 50 ppb. Some trace elements may have lower values, as in the Super Premium tier.

The methodology to test products that meet the projected specifications will need to be developed. Currently, an Inductively Coupled Plasma/Mass Spectrometer is the only instrument that is suitable for performing the analyzes with reasonable accuracy. With the expense of buying such an instrument at $300,000–$400,000 and expensive, highly qualified operating personnel, there will be a significantly higher cost of quality control for these purer products. Improvements in flameless atomic absorption may also assist in more accurate measurements of metals.

The development of processes to produce products of this purity has been, in large part, achieved by the Japanese chemical companies. Their approach has been to start with purer starting raw materials, upgrade the products through distillation in equipment with expensive inert metals and fluoropolymer materials of construction that flow through dedicated lines and filters to storage tanks, and then are packaged in cleanrooms.

Their overall approach to minimizing any source of contamination, whether in the process, the transfer of product, or the packaging materials, has been successful. The U.S. and European semiconductor industries will need to adopt this approach widely. There are some products from some U.S. vendors that are approaching the purity of the Premium grade already.

If the semiconductor industry in the late 1990s needs purity that is measured in the part per trillion range, different processing techniques will be required. Use of small scale vacuum or sub-boiling distillation techniques, which are used in laboratory-scale production, could be scaled up to larger-volume production. Some products could be purified by ion exchange techniques. Inherent is using these or other techniques will be the downstream problem of protecting the product's purity in transport to the fab point-of-use.

3.2.2. Particulates

Effects on Yield

As device dimensions become increasingly smaller, the issue of particulate contamination has gained increased emphasis. A direct relationship between the scale of integration for memory devices and the critical size of particles sufficient to cause the device to fail has been shown, as illustrated in Figure 3.1. This relationship is approximately 10% of the minimum feature (Brandt 1983). In the case of 256 Kbit DRAMS, for example, if the design rule is 1.4 μm, the diameter of the particles that must be removed is 0.14 μm or larger.

Particulates in acids and solvents arise from two sources—inherent and transferred. Inherent particles are in the as-received containers, sinks, and process equipment while transferred particles are generated during the etching or cleaning process itself from piping, filters, pumps, or shedding of the materials of construction. Additional sources of particulates are from workers handling the chemicals and from process equipment. The largest source,

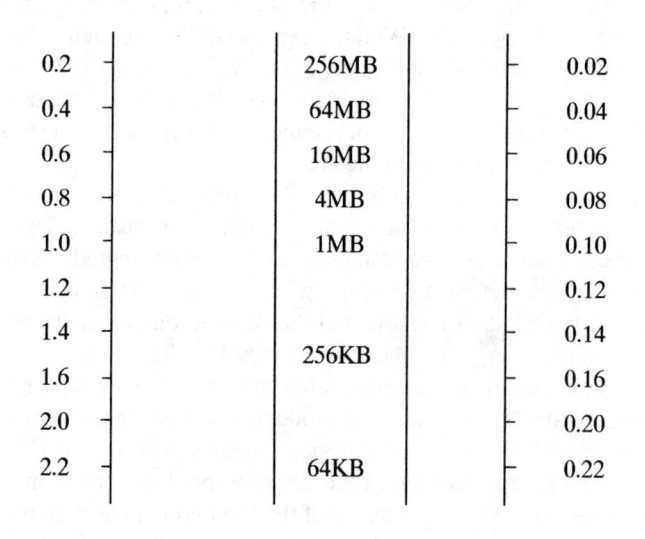

DESIGN RULE (μm)	MEMORY CHIP CAPACITY	CRITICAL PARTICLE SIZE (μm)
0.2	256MB	0.02
0.4	64MB	0.04
0.6	16MB	0.06
0.8	4MB	0.08
1.0	1MB	0.10
1.2		0.12
1.4	256KB	0.14
1.6		0.16
2.0		0.20
2.2	64KB	0.22

Figure 3.1. **Relationship between feature size and critical size of particles.**

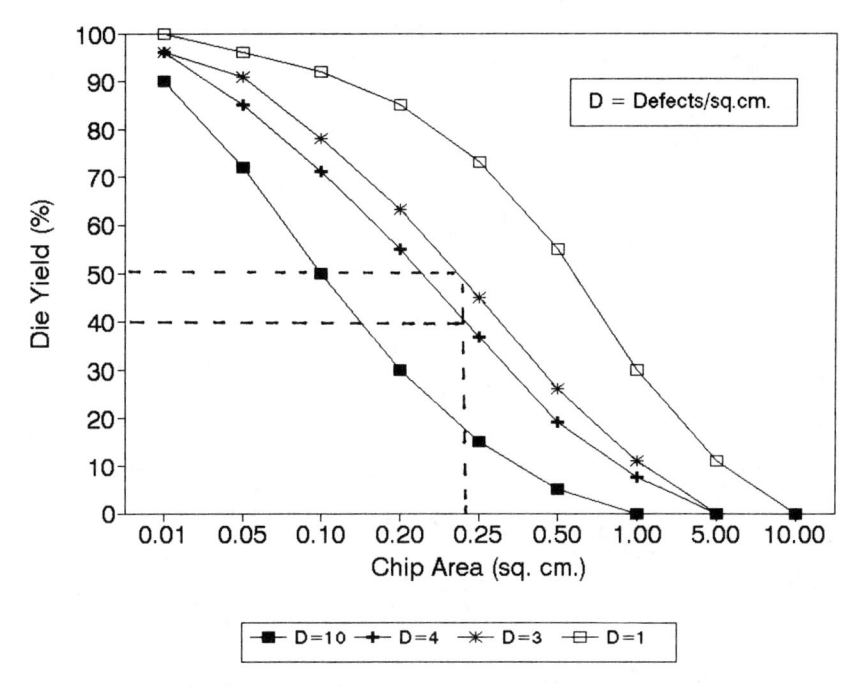

Figure 3.2. **Relationship between die yield, defect density and chip size.**

however, is from the wafer processing itself. The chemicals react with the processed wafer and generate particles, inorganic colloids, and residues.

The yield and reliability of semiconductor devices is a function of particulate contamination in all stages of fabrication. Random defects on photomasks are the major source of yield loss in semiconductor devices, with particulates accounting for nearly half of these defects. As shown in Figure 3.2, device yield is proportional to the size of an IC die, which has been increasing with each generation of devices. Toshiba's 4 Mbit DRAM, for example, is 5.4 mm × 15.1 mm. Die yield is, in turn, related to the cost per chip as illustrated in Figure 3.3. Thus, the reduction of defects generated by particulates in chemicals (as well as operator and equipment generated particles) will increase yields and revenues.

Particle removal is difficult because of the tenacity with which particles adhere to the wafer surface. The relative adhesive forces of glass spheres on a glass surface are as large as 58,000/sphere weight for a 10-µm diameter

34   LIQUID CHEMICALS

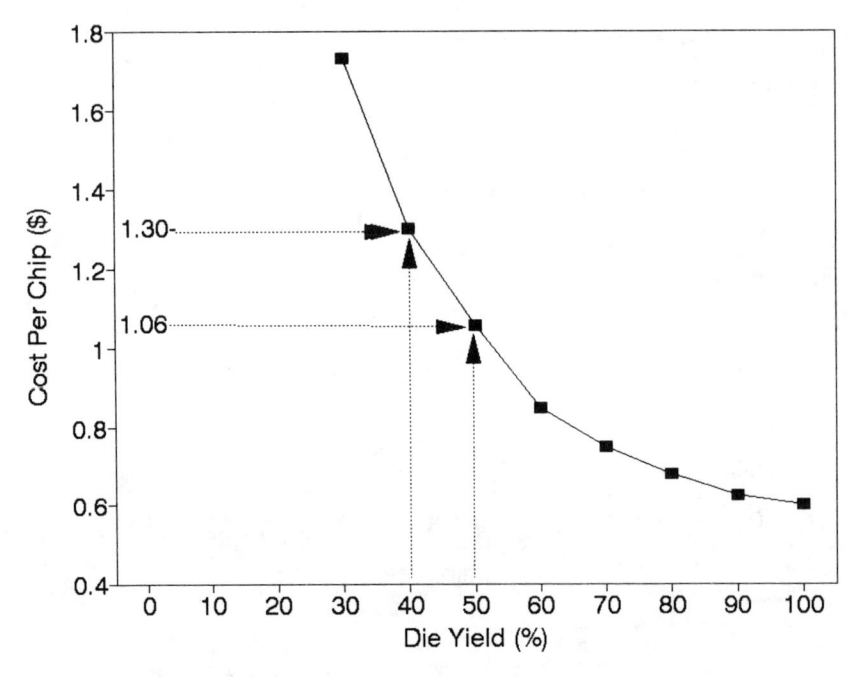

Figure 3.3. Relationship between die yield and cost per chip.

sphere and increase to 675,000/sphere weight for a 1-µm diameter sphere (Rechen 1985). Thus, complete removal is virtually impossible once µm-size particles adhere to a wafer surface. Two methods of removal are common, wet-chemical cleaning and high-pressure filtered nitrogen. Although these methods generally avoid the introduction of additional contamination, there is no significant removal of particles except for extremely dirty wafers.

Low-particulate chemical usage in the semiconductor industry is increasing. Low-particulate materials are available at prices two to three times the cost of Semi-grade chemicals. Further, advances in the industry incorporate the use of 55 gallon fluoropolymer drums for low-particulate chemicals. The price of chemicals purchased in this container is approximately 2.5 times less than purchased in bagged bottles.

Most of the chemical suppliers are addressing the particulate issue by filtering chemicals through 0.2 µm filters prior to bottling. This operation can

LIQUID CHEMICALS

be performed overnight, and the tank recirculated through the filter a number of times so that particulates are reduced. Although not bagged, the level of particulates can be as low as the low-particulate but at essentially no extra cost. Future trends will see greater usage of bagging of bottles sold in returnable, fully-enclosed cases, and then shrink wrapped.

Photoresist can become contaminated from a number of sources before arriving at the fab line (Long 1984):

- Container material
- Shipping and storage
- Operator cleanliness
- Spinner bowls
- Soft bake ovens

The primary source of contamination is the container. Glass bottles must be manufactured and sealed under expensive cleanroom conditions since small particles are held with sufficient tenacity that complete removal by washing the container is nearly impossible to achieve. Special manufacturing techniques are thus required. Poly bottles are worse, since greater static forces retain particles more strongly. Also, there is evidence of reaction of the poly bottles with caustic chemicals. The newest packaging container for positive resist is the pouch. Contact of the resist with air is virtually eliminated, and the multilayer construction facilitates high-purity fabrication.

Shipping and storage are an additional source of contaminants. If resist is not shipped in temperature-controlled vehicles, resist degradation and accelerated particulate formation results. Extended storage of positive resists can result in aging, in which the sensitizer degrades and precipitates out of the liquid. Refiltration is required if the resist is stored for 6 weeks or longer after the initial purification process. Contamination of photoresist in the fab line occurs most often at the dispensing system by the transfer of the dip tube from bottle to bottle. Use of the pouch reduces contamination by eliminating the need for the dip tube.

Particulate Removal Techniques

Filtration of chemicals, with filters placed as close to the process equipment as possible, is now the norm. Since the etching and cleaning process itself is a major source of particulates, these liquids must be continuously filtered. Because of the corrosive, caustic or dissolving capabilities of these liquids, point-of-use filters must be inert, self-wettable for ease of handling, and must not release fibers (shedding) (Tolliver 1984).

LIQUID CHEMICALS

Considerations in point-of-use liquid chemical filtration that are important in semiconductor manufacturing are:

- Compatibility
- Cost
- Efficiency
- Flow rate
- Maintenance
- Pore Size

To optimize the particle removal while minimizing the face velocity to less than 1 gal/min/ft^2 of membrane area, filters should have a large surface area. A variety of liquid chemical filters are commercially available. These are constructed of several different types of materials but generally have pore sizes of 0.2 µm, which, in principle, permit only particles of this and smaller size from penetration. However, larger particles do pass through. Material types include Teflon, polypropylene, polyvinylidine fluoride, polysulfones, and PTFE (Krygier 1986).

Although basically inert, all filtration devices contribute some level of contaminants into the chemicals due to leaching, whereby the filter is subjected to swelling and solvation by the chemicals. This leached material is generally less than 0.1% of the filter weight and is removed quickly in the filtration process. Stacked membrane filter construction offers lower levels of leachable material than do pleated filters.

There are several methods to remove particulates from wafer surfaces. Each have their advantages and disadvantages, as shown in Table 3.4 (Menon *et al.* 1989).

The most common methods are:

- Wet chemical cleaning is often performed in Teflon or polyethylene sinks that are either manually operated or automated via robotics on an overhead gantry.
- Brush scrubbing utilizes polypropylene, nylon, or mohair bristles to sweep off contamination from the wafer surface although the bristles of the brush do not come in contact with the wafer surface.
- Sonic cleaning employs ultrasonic agitation (cleaning action accomplished by bubble generation) or megasonic scrubbing (cleaning accomplished by high pressure waves).
- High pressure or centrifugal spray acid processors use a jet of fluid at up to 4000 psi to sweep off particles from the surface of the wafer.

Table 3.4. **Advantages and disadvantages of various cleaning methods.**

Cleaning Method	Advantages	Disadvantages
Wet Chemical	Removes metal ions and soluble impurities	Not good for particle removal Requires water rinsing and drying Known to add particle to water
Scrubbing	Removes large particles (> 1 micron) effectively Suitable for cleaning hydrophobic wafer surfaces	Unsuitable for removing submicron particles from patterned wafers Requires regular maintenance Could damage wafers
Pressurized fluid jets	Removes small particles from patterned wafers	Possibility of wafer damage Static buildup
Ultrasonics	Removes small particles effectively	Wafer damage possible Control of cavitation intensity difficult Unable to use reactive solutions
Megasonics	Removes both soluble and insoluble contaminants Removes small particles effectively Can be used with chemical cleaning solutions Less likelihood of wafer damage	Contamination from chemical solutions possible Wafer rinsing and drying required
Strippable polymer tape	Capable of removing both soluble and insoluble impurities Convenient for wafer storage and transport Dry technique	Possibility of depositing polymeric residue Effectiveness not demonstrated adequately
UV/ozone	Removes organic impurities Dry technique	Not proven for particle removal Possibility of wafer damage

Researchers at Mitsubishi's LSI R&D Laboratory at Itami City have explored the use of ice scrubbing to remove particles (Ohmori 1989). In this method, clean liquid nitrogen is used to form ice particles when ultrapure deionized water is sprayed into a chamber. The particles, measuring between 30 and 300 µm, are sprayed onto the wafer via a nitrogen-gas carrier. The removal rate of standard 0.322-µm polystyrene particles from wafer surfaces for this method and other common removal methods is shown below:

Cleaning method	Removal rate (%)
Ice scrubber	95.0
Megasonic	95.0
Brush scrubber	87.4
High-pressure water scrubber	84.4
Ultrasonic	83.9
Dry ice scrubber	68.9

Stacked membrane filter construction offers lower levels of leachable material than do pleated filters. This type of filter has no woven downstream supports or flexible components to generate particles. A fluorocarbon membrane is rigidly supported on a polysulfone disc, which makes the filters suitable for use with a wide range of chemicals and resists to 0.2 µm retention. All-Teflon stacked disc membrane filters are available for corrosive fluid filtration with retention ratings of 0.1 and 0.2 µm and flow rates of 7 gallons per minute.

Shown in Table 3.5 is a list of the compatibilities between processing chemicals and the materials of construction of the filters.

One of the major problems that filter manufacturers experience is the proving of their filters. Recently, microprocessor-based automatic filter test equipment has been developed for testing the integrity of membrane filters in critical applications (Domnick Hunter Filters, Co. Durham, UK). The instrument is portable with a handheld control console for manual or automatic operation and reprogramming, providing a printout of bubble point and pressure decay test values.

Particle Monitoring

Particle monitoring of liquids is performed by two methods; light scattering and ultra-sound—both with a sensitivity of 0.4 µm (Knollenberg 1983) (Cheung and Roberge 1986). Particulate levels up to 3500 per ml of liquid at flow rates of 100 ml/min can be measured. Although in-situ monitoring of corrosive chemicals such as hydrochloric and sulfuric acids has been demonstrated by various suppliers, most liquid particle monitoring systems in use in fab lines today are applied to DI water.

In current semiconductor manufacturing, chemicals are changed or replenished in an etching or cleaning station on either a time or throughput basis. Certain solutions such as the Piranha etch are changed every 4 to 8 hours because of peroxide degradation regardless of the number of wafers processed. Oxide etchants, on the other hand, are generally changed after a specific number of wafers are etched because of changes in acid concentration.

Table 3.5. Chemical and material compatibility.*

Chemicals	Concentration (percentage)	PP	PVDF	Nylon	PFA	Polysolfone
Acids						
Acetic	80	L	R	NR	R	R
Hydrochloric	35	L	R	NR	R	R
Hydrofluoric	60	R	R	NR	R	L
Nitric	70	NR	L	NR	R	L
Perchloric	50	L	R	NR	R	NR
Phosphoric	80	R	R	NR	R	R
Sulfuric	98	NR	L	NR	R	NR
Bases						
Ammonium hydroxide	28	R	R	R	R	R
Potassium hydroxide	50	R	R	L	R	R
Sodium hydroxide	70	L	R	L	R	R
Other chemicals						
Acetone	100	R	L	R	R	NR
Ammonium fluoride	25	R	R	R	R	R
Ethyl acetate	–	R	NR	R	R	NR
Hydrazine	–	L	NR	NR	R	NR
Hydrogen peroxide	90	R	R	NR	R	R
Methanol	–	R	R	NR	R	R
2-Propanol	–	R	R	L	R	R
Xylene	–	NR	R	R	R	NR
Mixtures						
Aqua regia	–	NR	R	NR	R	NR
Buffered HF	–	R	R	NR	R	L

*R, recommended; L, limited; NR, not recommended; (based on ambient temperature exposure).

LIQUID CHEMICALS

Nevertheless, particulate levels are not considered in either of these operations. Real time, automatic monitoring of liquid particles by instruments that are integrated into wet chemical etching and cleaning systems are commercially available but used by relatively few semiconductor manufacturers. This situation is changing rapidly, as semiconductor companies have begun to recognize the potential savings by utilizing this equipment. With high throughput levels, sampling the bath for subsequent measurement in the QA lab leads to time lag inaccuracies. Also, most manufacturers have not invested the necessary time to determine the cost effective point to change chemicals. Point-of-use monitoring could be utilized to determine if particulate levels are too high during processing prior to regular maintenance schedules. Point-of-use monitoring could also be used to determine the integrity of point-of-use filters in recirculation systems.

Trends for Purity—Particle Control

In the first half of the 1990s, we project a two specification system for particles:

- *Super Premium.* Less than 50 particles greater than 0.2 µm per milliliter of product.

- *Premium.* Less than 10 particles greater than 0.5 µm per milliliter of product.

We project that the above standards will be likely based on the work being done at Sematech and the SEMI standards committee on chemicals. Products that meet the projected Premium grade will not be difficult to produce using current filtration technology, nor will the Super Premium grade be that difficult to produce. However, the problem for the suppliers is twofold:

- Packaging the product for delivery to the customer so that the particle specification is maintained at the time of usage will be difficult. The solution is a movement to deliver the products in large, recyclable containers.

- Even if the packaging/delivery problem is solved economically, will tighter point-of-use specifications force the fabs to conduct the final particle control steps in the fab facility? If so, chemical companies only need to supply a prefiltered product, probably at current or slightly improved specifications.

LIQUID CHEMICALS

Table 3.6. **Common gallon containers for standard chemicals.**

	Glass*	Poly**	Fluoropolymer†
Sulfuric	o	o	o
Nitric	o	np	o
Hydrochloric	o	o	o
Acetic	o	o	o
Ammonium hydroxide	o	o	o
Phosphoric		o	o
Hydrofluoric		o	o
Ammonium fluoride		o	o
Buffered oxide etchants		o	o
Isopropyl alcohol	o	o	o
Methanol	o	o	o
Hydrogen peroxide		o	o
Xylene	A	np	o
Trichloroethylene	o	np	o
Trichloroethane	o	np	o
Acetone	A	np	o
n-Butyl acetate	A	np	
Sodium hydroxide		o	o
Potassium hydroxide		o	o
Toluene	A	np	

*A, amber
**np, not permitted
†, limited due to high cost

3.3. DELIVERY ALTERNATIVES

3.3.1. Bottles

Glass and Polyethylene

The majority of chemicals used in semiconductor fabrication are supplied in glass or polyethylene gallon containers. In fact, polyethylene is commonly used in most of the processing chemicals. When there is a choice of container for a given chemical, polyethylene is preferred by the user, primarily for cost and safety reasons. For example, more than 70% of sulfuric acid available in gallon containers is supplied in polyethylene. For isopropyl alcohol, this level increases to 98%, with only Hewlett Packard (among large companies) purchasing IPA in glass.

Shown in Table 3.6 is the alternative bottling available for the standard chemicals used in IC manufacturing.

New Materials

From the information discussed above, it appears that glass bottles are superior to polyethylene in terms of resistance to particulates and impurities. One of the drawbacks with glass bottles concerns cleanroom operator safety, governed by OSHA, and transportation safety, governed by the DOT. An impact resistant, thermo plastic coating has been developed by Wheaton Safety Container Company (Mays Landing, NJ) to protect glass from breakage and damage during handling and transit. This "Second Skin" also has a non-slip feature that aids in pickup by remote manipulators whether wet or dry. This bottle, however, has met with great resistance by the purchaser due to the significantly higher cost charged by the chemical supplier of $2–$3 per container over the equivalent standard glass bottle. Other materials, particularly fluoropolymer, have been introduced. However, the price is prohibitive in throw-away, gallon-sized containers.

3.3.2. Drums

Steel and Polyethylene

Shown in Table 3.7 are the most common drum liners for semiconductor processing chemicals.

The most commonly used drum today is polyethylene. The most common sizes are 5 gallon (or 20 liter) and the classic 55 gallon drum. The use of the smaller size drum has been driven by the per pound cost component of the drum versus bottles and the lesser quantity of units to handle and dispose, Most users of the smaller drums have adopted some sort of dispensing system so that the chemicals do not need to be poured directly from the container; therefore improving worker safety.

Steel drums are primarily utilized in solvent service because of the possibility of extractables such as antioxidants or a softening of the polyethylene. Steel drums with polyethylene liners are used where added safety is required by either international or DOT shipping regulations.

Fluoropolymer

The need for a container that better protects the purity of the product has spawned the introduction of the fluoropolymer drum. The drum is composed of a fluoropolymer liner housed in a steel or plastic overpack. Introduced in the late 1980s by Fluoroware and RMB Products, the container was not readily accepted due to the high cost of $1500 (with carbon steel overpack) and the typical start-up problems associated with new product. The cost of the drum could naturally make recycling mandatory, and the logistics proved to be tough for supplier and customers to learn.

Table 3.7. **Common 55-gallon drum liners for semiconductor processing chemicals.**

	Poly	Poly-lined steel	S.S.	Fluoropolymer
Sulfuric	o			
Nitric				o
Hydrochloric	o			
Acetic	o			
Ammonium hydroxide	o			
Phosphoric	o			
Hydrofluoric		o		o
Ammonium fluoride		o		
BOE	o	o		o
Hydrogen peroxide	o			
Isopropyl alcohol	o		o	
Methanol	o		o	
Xylene			o	
Acetone			o	
n-Butyl acetate			o	

Note. Also generally applies to 5, 15, and 30 gallon sizes.

Fluoroware is currently the only active supplier. The 1990s will see rapid growth in the use of returnable fluoropolymer containers. Fluoroware has several models now available including a pressurized design.

The Fluoropure Quick Connect System, which permits easy connection of the drum to a dispensing system, has also contributed to the acceptance of the drum-dispensing system of chemical delivery to the fab. The Quick Connect System can also be used with all polyethylene drums.

Returnable Drum Economics

The initial resistance to returnable fluoropolymer drums was over the method of cost recovery for the expensive capital cost. Customers expected the supplier to purchase the drums and provide the product at little increase in cost versus conventional one-way drums. The requirements for a deposit and all the paper work were very troublesome to the customers.

The basic economics were initially uncertain because the useful life of the drums was unproven. Looked at in terms of current pricing for a 55 gallon poly overpack model at $1900, a drum can hold 750 pounds of sulfuric acid or 500 pounds of hydrofluoric acid. This means that the capital cost of $2.50 to

$3.80 per pound has to be amortized over a number of trips. If 40 to 50 trips can be made, then the cost amortization of a fluoropolymer drum is approximately equivalent to a polyethylene drum. However, this ignores two important factors:

- How long would it take to achieve 40–50 cycles? If the shipping time, filling time, and inventory residence time at the supplier and customer are taken into account, a drum would typically achieve only 4–5 trips per year. Implying a 10-year cost recovery period, this is not attractive to customers.

- The cycle time length means that a large number of drums have to be purchased to provide the needed level of service. The total capital cost of providing the drums is a major issue.

The supplier concerns over the cost of fluoropolymer has been eased somewhat by the recognition of the customers that they will have to purchase the drums (which several companies have done), or pay a rapid amortization factor in the price of the chemical that is guaranteed in some fashion to cover the capital cost. Also, the practice of using fluoropolymer returnable drums has been limited to critical products such as nitric acid, etchants, and hydrofluoric acid by some customers. Other product can be shipped in all polyethylene returnable containers.

Olin Hunt is the only supplier to introduce electronically scanned bar coding to ensure precise tracking of containers.

Suppliers

General, Ashland Chemical, and Olin Hunt are front runners in fluoropolymer/poly returnable containers. Baker currently has a minor effort, but is ramping up its program.

3.3.3. Minibulk Containers

With the acceptance of returnable drums proceeding, the logical progression is to even larger containers; but as an intermediate to the existing bulk truck load capability established long ago. Various other industries have used minibulk containers that generally range in size from 200 to 300 gallons. The practical limiting factors are a convenient size and weight that can be transported by a truck making deliveries of palletized bottles and moved about by a forklift.

Minibulk containers are constructed of polyethylene or fluoropolymer usually surrounded by a protective steel mesh cage.

3.3.4. Bulk Truck Trailers

The semiconductor industry's larger companies have often installed facilities to receive chemicals by bulk delivery. Most of the major suppliers have fleets of bulk trailers to deliver the high volume products. Deliveries are of quantities that weigh the net load limits of the trailer, which is typically 45,000 pounds.

The lining of the tank compartment(s) has been typically glass for acids and stainless steel for solvents. Hydrofluoric acid and ammonium fluoride have had rubber-lined tanks. The concerns over purity have caused a movement of using fluoropolymer lined tanks on most new or refurbished trailers put in service in the past five years. The predominant lining material is either PVDF or PTFE. The lining and cost of a new trailer is on the order of $100,000.

3.4. IN-PLANT DISPENSING AND DISTRIBUTION

3.4.1. Bottles in the Cleanroom

Most chemicals are bottled by the supplier and shipped to the manufacturer. They are typically stored outside the fab area until needed (Dillenbeck 1991), and transported by push cart into the line where they are poured into cleaning or etching stations by the operator. There are several disadvantages in this practice:

- There is an increased cost of chemicals due to costs of the bottle, as well as added supplier expenses for bottling, labeling, handling, and packing (cardboard boxes, foam packs).
- Fab operators must handle individual bottles in terms of unpacking and clean-washing prior to entry into the clean room, as well as transporting them from a storage area in a safety carrying cart or individual holders. The transport of acids within the facility is often limited to specially designed pass-through cabinets and extra wide corridors that have to meet rigorous fire code specifications. During these procedures, and when handling and pouring gallon containers into waist-level tanks, operator safety is compromised in terms of breakage, splashing, and noxious fumes.
- There is an increased chance of operator error in the handling of thousands of bottles each week. This would increase the chance of process contamination.
- Empty bottle disposal must comply with local and state ordinances that require triple rinsing and tagging of each individual bottle to assure re-

moval of residuals before crushing and discarding. This has been estimated to amount to over a dollar per bottle in costs directly associated with disposal labor.
- Particulate contamination is increased from shedding of containers as well as from airborne, operator-generated particles.

3.4.2. On-site Packaging

Point-of-use dispensing into individual containers, and central piping systems into the fab area from bulk containers are currently two prevalent methods of bulk dispensing. This trend eliminates the problems associated with handling chemicals packaged in bottles. Bulk delivery systems fabricated from high purity fluoropolymer components are becoming wide spread in the Japan, but limited to large facilities in the U.S., in which acids are received in trucks.

Recycling bottles only makes economic sense within a fab site or small geographic range. It should be noted that Japanese fabs receive product in recycled bottles, but they pay higher prices for that service even though the economics are supported by its widescale adoption.

In the U.S., the primary proponents of this system are Texas Instruments and Motorola. Motorola at Phoenix recently installed a multi-million dollar chemical mixing and bottling facility.

3.4.3. Direct Dispensing to Point of Use

In the review of chemical systems, fab operators have seen both the need to eliminate bottles from the fab and the advantages of on-site bulk-to-bottle facilities. The next step was to bring the chemicals straight to point-of-use in a pipe system. The ability to bring the chemical to the point-of-use was also encouraged by the adoption of equipment that used the chemicals in an automated fashion, such as spray etchers and rinse dryers (Hashimoto *et al.* 1989).

The Japanese first provided an integrated chemical dispense system that filtered and piped chemicals to the point-of-use station. Sumitomo and Kanto provide such systems to customers.

In the U.S., FSI International, Mega Systems, and Systems Chemistry have introduced chemical dispensing systems. Several other equipment manufacturers offer similar systems, but on a more customized design basis. The basic advantage of the system is the elimination of bottles by a linkage from a bulk storage tank (or returnable drum or minibulk container) to the point-of-use station by means of pumps, pipes, filters, valves, and a control system.

LIQUID CHEMICALS

The other advantages of the chemical dispensing systems are:

- Re-engineered design as opposed to custom in-house approach
- Modular components that are easily expanded
- Flexibility of chemical supply sources—drums, minibulk, or bulk
- Economical for even small volume chemicals
- Vendor installation and service support

New drums and newly designed dispensing accessories and new air operated, corrosion resistant pumps make remote dispensing of chemicals into clean rooms a much more feasible alternative. Satellite dispensing systems and a variety of chemical dispensing carts are available today as well.

3.5. CHEMICAL REPROCESSORS

Chemical reprocessors are recently developed systems that are utilized for recycling chemicals used in semiconductor processing (Davison and Hoffman 1987). The concept of reprocessing chemicals used in the semiconductor fab is a natural idea that has been accelerated in its development by environmental concerns and the recent commercial availability from two companies—Athens Corporation and Alameda Instruments. There has been a natural economic progression of reprocessing the easy chemicals first. The types of chemicals used in a fab can be looked at as pure chemicals and mixtures. Product such as solvents used in rinsing and cleaning steps have been reprocessed for a long time, as the product was collected separately from other chemicals, then sold to solvent reprocessing companies that redistilled the waste solvents for sale to other end users.

With acid wastes, the disposal is principally accomplished by sewering after neutralization with sodium hydroxide. However, if the acid waste stream contains fluorides, it must be handled differently. Either the fluoride can be precipitated through introduction of calcium (limestone or sea shells) and the precipitate sent to a hazardous waste site, or the waste stream hauled away for deep well injection or other permitted disposal.

3.5.1. Piranha Reprocessors

Background

The piranha bath, a combination of sulfuric acid and hydrogen peroxide in 6–10 parts sulfuric to 1 part hydrogen peroxide, is a ubiquitous fixture in the semiconductor industry. Since the piranha bath is used to remove photoresist from the wafer surface, it causes sulfuric acid to be the highest volume chemical to be used by the industry and therefore one of the bigger disposal concerns.

The use of piranha baths is fundamental in U.S. semiconductor technology. The use of piranha is, in large part, due to its low cost versus other wet chemicals and significantly lower capital costs versus plasma (ashing) stripping.

Suppliers

Two companies market piranha reprocessors—Athens Corporation and Alameda Instruments. The piranha reprocessor unit is based on a continuous two-step distillation of a stream from the bath to first remove water and volatile components and a second distillation to remove high boiling components and heavy metals. Electrolytically-generated peroxysulfuric acid is added as the oxidant (replacing the traditional hydrogen peroxide or ammonium persulfate) to the purified sulfuric acid and the mixture returned to the bath.

The Athens unit uses atmospheric distillation in the final distillation at an operating temperature of 300–350°C. The Alameda Instruments unit uses a vacuum distillation at 40 mm Hg at an operating temperature of approximately 190°C. Both companies' units cost in the range of $250,000–$400,000 depending on specific features.

Benefits of Piranha Reprocessors

As part of the recycling processing, several benefits can be obtained (Jones, 1987):

- Reduction of particles per wafer in prediffusion clean to zero as shown in Figure 3.4
- Reduction of trace impurities in piranha due to the two-step distillation to a level of 100 ppb or lower
- Reduction in piranha consumption by 90–95% compared to a traditional piranha bath
- Improved process control by:
  — Controlling oxidizing
  — Counting particles on-line
  — Measuring trace impurities on-line
- Similar reduction in environmental disposal costs of spent material
- Consistent piranha concentration and purity versus standard bath that degrades over time

LIQUID CHEMICALS

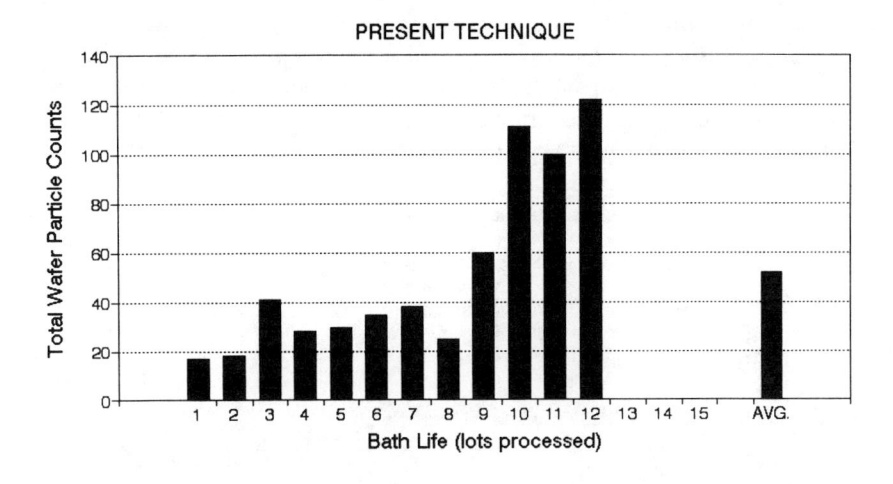

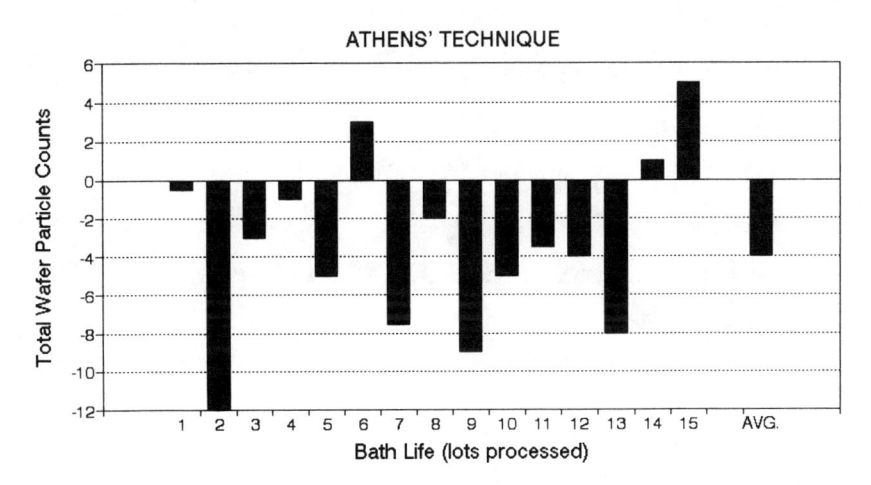

PRESENT TECHNIQUE

ATHENS' TECHNIQUE

Figure 3.4. Particle count reduction with chemical reprocessors.

Piranha Reprocessor Economics

Both reprocessor companies have provided economic evaluations of their equipment based on operating savings achieved versus the use of bottled chemicals. Some studies indicate that a significant device yield improvement is derived from the reduced particles and lower trace element attainable in a reprocessor piranha bath. Given the high cost of the equipment, the justification for their purchase should include the improvement in yield obtained from the purer product that the reprocessors produce. The reduction of chemical purchases and environmental disposal costs are secondary added value.

The cost analysis/payback is given in Table 3.8. Based on incoming chemicals and operating conditions, a savings of $9,471 per month for semi-grade chemicals and $46,324 per month for megabit-grade chemicals per reprocessor can be achieved.

The Athens' Piranha Piranha system is currently at:

- AMD (Austin, TX)—CMOS E^2/EPROM
- AMD (Sunnyvale, CA)—0.6 µm CMOS
- Digital Equipment
- Harris (Mountaintop, PA)—Power CMOS/ASIC
- IBM (E. Fishkill, NY)—multiple units—Bipolar memory/logic
- Intel—multiple units
- NEC (Sagamihara, Japan)—DRAM
- Samsung (Suwon, Korea)—DRAM
- Sematech (Austin, TX)—DRAM/SRAM
- SGS-Thompson (Carrollton, TX)—CMOS ASIC/µP
- Tech Semiconductor (Singapore)
- Texas Instruments (Lubbock, TX)—CMOS EPROM
- Toshiba (Iwate, Japan)—CMOS ASIC/µP

3.5.2. Hydrofluoric Acid Reprocessors

Both Athens and Alameda Instruments have developed reprocessors for hydrofluoric acid, but use different technology. Alameda (whose project is currently on hold) uses a distillation approach that is essentially the same as their piranha reprocessor. Athens has adopted an ion exchange technology, which is effective in reprocessing dilute hydrofluoric acid streams of up to 10% strength. The Athens unit produces a product that is no stronger than the initial waste stream, unless it is augmented with purchased high strength (49%) material. The Alameda unit will concentrate the waste stream up to 37% strength. As the inherent cost of disposing of fluoride wastes is greater than that of sulfuric acid, these units should have a large market potential.

Table 3.8. Cost analysis/payback of reprocessor.

		Semigrade	Megabit	Athens
1.	Chemical Consumption			
	• Sulfuric acid (gal./mo)	2,167	2.167	65
	• Hydrogen peroxide (gal./mo.)			
	—Initial charge	477	477	0
	—Spiking	1,140	1,140	0
	Total H_2O_2	1,617	1,617	0
	Total Piranha Solution	3,784	3,784	65
2.	Chemical Costs			
	• Sulfuric acid	$17,250	$26,849	$805
	• Hydrogen peroxide	$6,703	$33,957	$0
	Total Chemical Costs	$23,953	$60,806	$805
3.	Chemical/Bottle Disposal			
	• Neutralization/dilution $(1.00/gal.)	$3,784	$3,784	$65
	• Bottle/rinsing/disposal ($1.25/gal.)	$4,730	$4,740	$0
	Total Disposal	$8,514	$8,514	$65
4.	Operational Costs			
	• Fab handling (1/shift, 3 shifts/day)	$7,740		$0
	• Chemical maintenance (purchasing, receiving, storage, QC)	$4,267		$1,799
	• Utilities/facilities differential	–		$3,000
	• Service contract	–		$8,000
	• Miscellaneous parts	–		$3,000
	• System depreciation ($1M)	–		$16,667
	• Installation ($100K)	–		$1667
	Total Operations	$12,007		$34,133
Grand Total				
	Reprocessor Savings	$44,474	$81,327	$35,003
	Monthly Savings	$9,471	$46,324	

The Athens' HF system is currently at:

- IBM (East Fishkill, NY)—Bipolar memory/logic
- Sematech (Austin, TX)—DRAM/SRAM
- Texas Instruments (Dallas, TX)—Multiple units—DRAM
- Texas Instruments (Dallas, TX)—Multiple units—various

3.5.3. Strategic Issues and Future Developments

Much of the improvement derived from the use of reprocessors results from essentially moving the chemical plant into the fab, eliminating most sources of product contamination from packaging and transferring the chemicals. However, the notion of having a "chemical factory" within the fab is a concern to many semiconductor companies for environmental reasons.

The impact of piranha reprocessors on the chemical suppliers, if the reprocessors were widely adopted, would be devastating, due to the 90–95% reduction of sulfuric acid demand. The recent price cuts in premium sulfuric acid grades relative to other premium chemicals may be partially explained as an attempt to reduce the economic incentive to install piranha reprocessors.

The extension of the reprocessor distillation and ion exchange technology to other acids and chemical mixtures is possible. The main considerations are the degree to which the chemical is contaminated in use, and the economic ability to segregate its waste stream. Buffered oxide etchants, which are mixtures of ammonium fluoride and hydrofluoric acid, can be reprocessed at high volumes into hydrofluoric acid by a modified hydrofluoric acid reprocessor. Other etchant products containing nitric, acetic, and phosphoric acids will be difficult to reprocess.

Athens has announced several programs with chemical suppliers:

- Air Products will market the chemical purifier. Air Products will own the equipment, and install it at the customer's site. The company will also perform all maintenance and service under a monthly service fee for the customer. Four systems were installed by the end of 1991.
- Sumitomo Chemical will market the system in Japan.
- J.T. Baker, Texas Instruments, and Athens Chemical have begun a program whereby surplus 1 ppb-grade HF from the reprocessor will be bottled by TI and sold by J.T. Baker to semiconductor firms. The product is known as Athens Acid. Already, TI has stopped using 1 ppb HF from its Japanese source and is using the Athens Acid.

Similar programs will be developed for purified sulfuric acid.

3.6. SUPPLIER RATIONALIZATION

The purchasing practices of the semiconductor industry, the need for greater purity and consistency, and vendor-customer partnering is going to reduce the number of major suppliers of process chemicals active in supplying the high-purity products. The process has been under way since the 1985 semiconductor industry slump. Several suppliers have been eliminated through acquisition and withdrawal. The only significant entrant has been Olin, through the purchase of Hipure and subsequent significant investment in developing that business.

The past purchasing practices of the semiconductor industry were not focused on quality issues. Process chemicals were bought from the lowest bidders from a large pool of qualified vendors. Customers would often have four or five qualified vendors. Request bids covering supply periods of three months to a year were typical. Vendor qualification procedures ranged from reliance on simple in-house testing against SEMI specifications to more extensive split-lot comparisons on the production line.

This resulted in fabs buying the full range of process chemicals typically from four to five suppliers, with some fabs doing business with up to a dozen suppliers. These practices were costly and inefficient. Under this system, vendors were focused on providing minimal purity against specifications to eke out a profit.

The pressure of the device manufacturers to reduce variability in their manufacturing process is impacting the chemical suppliers in the U.S. through customer-vendor rationalization programs. Many of the larger device manufacturers have initiated formal programs to reduce the number of qualified chemical suppliers to what they believe is a manageable number. The goal of a consistent product suggests a single supplier. However, U.S. business culture dictates having second and even third sources for a given product. The need for a backup supplier to a primary source will enforce a pattern of having two to three qualified suppliers for a given product.

While the customers are reducing the numbers, they are also improving the vendor qualification procedures. The new rigorous qualification procedures include extensive review of the following areas:

- Analytical facilities and procedures
- Statistical process control programs
- Statistical quality control programs
- Manufacturing processes and equipment
- Deionized water facilities

- Packaging facilities—cleanroom capability
- Facility improvement programs
- R&D programs
- Corporate commitment

These reviews are accomplished by completion of extensive questionnaires, which are followed by facility audits.

The effect of these changes in purchase practices on the suppliers has been dramatic. Foremost is the ongoing squeezing out of the weaker suppliers that cannot make the investment in facilities and people to provide better product and service. The rationalization process has caused some suppliers to reduce their emphasis on the semiconductor industry. The market appears to need only three major suppliers that will provide the higher grades of product in advanced packaging. It is impossible to predict what three companies in each of the three categories discussed in this report will emerge as the suppliers of the 1990s.

3.7. STATISTICAL QUALITY CONTROL

Most chemicals sold today for semiconductor manufacturing are sold to specifications set forth by the chemical standards committee of the Semiconductor Equipment and Materials Institute (SEMI). These standards and methods of analysis have been developed over the last years by a qualified group of people from various disciplines within the chemical industry and the semiconductor industry.

The current SEMI specifications are available through SEMI, which publishes them in the Semi International Standards Volumes, Volume 1—Chemicals. There is a $75.00 fee for the single volume; the complete 5 volume set covering other standards is $200.00. The process chemical specifications include a particle specification for the first time.

The committee is presently reviewing all standards and methods to make sure they meet the needs of today's manufacturing requirements. Users are encouraged to communicate with SEMI regarding their company's specifications. It is advantageous to suppliers and users to adopt uniform specifications whenever possible.

Semiconductor manufacturers are expressing increasing concern that the present SEMI specs are inadequate. First, chemicals supplied by U.S. manufacturers are generally an order of magnitude cleaner than the SEMI label designation. Second, the SEMI tolerance levels are too large so that high lot-to-lot variations are not acceptable, particularly in advanced device fabs.

Most users continue to ask suppliers for higher and higher chemical purity. At the present time the supplier specifications represent product purity that is satisfactory for most manufacturing processes and that is commercially available at prices that are acceptable to users.

It is well known that the yields in Japanese IC manufacturing facilities are higher than in the United States. This is related to two major factors; the increased use of automation in Japan, and the decreased contamination levels due to the integrity of their Class 1 and 10 cleanrooms, their attitude to quality, and their processing chemicals. U.S. manufacturers are adopting these practices in order to remain competitive with the Japanese.

A major effort must be directed towards chemical manufacturers to supply chemicals with as high a purity and as low a particulate level as possible. This must also be achieved at as low a cost as possible. General's low-particulate product was not enthusiastically accepted by the semiconductor community because of the high cost of the product. Indications are that their Class 10 line will.

It is ironic in this industry that efforts to supply the highest purity products have not been standardized. For example, Class 10 cleanroom air contains 0.00035 particles greater than 0.2 µm per milliliter. In contrast, standard-grade processing chemicals contain 100–100,000 particles greater than 1.0 microns per milliliter. Even low-particulate chemicals can contain 10–500 particles per milliliter.

Higher purity chemicals could be offered but would require additional processing such as distillation, chemical treatment or even drastic modifications in the basic manufacturing procedures. Many basic chemical producers have already made improvements in their manufacturing facilities to accommodate the semiconductor industry and its chemical suppliers. Individual company specifications make it necessary for suppliers to do lot selection or extra analytical work and this can escalate the selling price of the chemicals. Also, it causes shipping delays since the supplier usually cannot ship his regular product from existing inventory.

Obviously these basic changes would change the cost structure and ultimately increase the selling price to users. These price increases could be dramatic.

The following are important comments on the three areas that semiconductor customers concentrate on when evaluating the purity of a supplier's product:

Assay and Related Items

The specifications for assay and related items have not been the subject of significant tightening in the past ten years. Certain products, particularly etchants, have had the analytical range for assay of key components reduced by important customers in their purchasing specifications. Items such as color, residue, water content, and acidity have been stable.

The major issue, as in other areas, is consistency. Purchasers are expecting the manufacturers to implement statistical process control/statistical quality control (SPC/SQC) programs as they have done in their own manufacturing processes. Advanced technology semiconductor companies are demanding specifications that not only specify tighter ranges than in the past, but the fluctuations within the range must be an acceptance criterion.

Furthermore, some of these customers are expecting that historical information on the production process and recent lot information (whether or not they bought any of the lots) be provided or made available upon request.

The establishment of SPC/SQC programs and the data collection effort are costly to the producer in terms of direct cost. A more subtle issue is what the manufacturers do with the product that does not meet the newer specifications.

Trace Elements

The chemical producers have been under tremendous pressure by the customers to reduce the trace element content of semiconductor chemicals as semiconductor device geometries have reached the submicron level. The level of trace element impurity was reduced by the Japanese chemical companies in the early 1980s. The methods the Japanese use primarily center around higher feedstock purity and better materials of construction throughout purification and packaging. There is little evidence that superior purification technologies are being practiced. U.S. and European companies have been following with improvements of their own, making significant strides in the last three years.

Due to the number of elements specified, one concept used to reduce the complexity when discussing specifications has been to add all the values for the trace elements to arrive at a total. Most discussions of the trace elements now start with the question "what is the total trace element content?," which is answered by adding the analytical results. The SEMI specifications generally cover 23 trace elements with maxima of 0.5 to 1.0 ppm. Therefore, is one adds the trace element maximum values, the various products will have a range of trace elements of 11 to 18 ppm.

A total content of 11 to 18 ppm may have been acceptable when producing semiconductor devices in the 1970s and early 1980s. However, purity requirements in the 1990s are in the range of 0.1 to 2 ppm.

Customers that are making the most advanced devices will require trace element purity in the parts per billion (ppb) range. Typical total values are requested to be below 100 ppb, with no single element above 10 ppb. **The Japanese have premium grades that can meet these specifications**. The U.S. producers are beginning to provide premium grades that match the Japanese purity in some products, particularly acids.

Sematech, the U.S. industry consortium organized in 1988 to develop new manufacturing techniques to respond to the Japanese challenge, has set probably the most stringent new specifications for wet chemicals. The specifications are proprietary, but industry speculation is that they are in the vicinity of a total maximum of less than 25 ppb total. The primary suppliers are thought to be Ashland Chemical and General Chemical.

If specifications are established at a 1 ppb level or below, new methods of purification will need to be implemented by the chemical producers. The techniques may not be significantly different from current practice, but the materials of construction will differ and the throughput will likely be reduced. Evidence of this approach to higher purity can be seen in its implementation in the piranha bath reprocessors. One other technique that is known to work on a small scale is sub-boiling distillation.

Any manufacturer that attacks the trace element problem aggressively will develop a significant market advantage for selling to the advanced technology semiconductor companies.

Particles

During the 1980s, as semiconductor device geometries were reduced, the concern with insoluble particle matter in the range of 5 to 0.5 µm in the wet chemicals became greater. These particles, which are not visible but could be measured quantitatively by optical microscopy or laser particle counters, were known to cause device failure. Several chemical companies, as previously mentioned, introduced premium grades for particle control in the late 1970s that had specifications of less than 10 particles measuring 1.0 µm or greater in size per ml. More vendors followed with similarly specified product grades in the mid-1980s.

Currently, most U.S., Japanese, and European chemical suppliers produce a standard grade product meeting the proposed SEMI particle specifications. These suppliers generally offer one or two premium grades that are specified at the 0.5 µm level.

The particle count reductions achieved during the 1980s were achieved by better filters with capability to capture particles down to 0.2 and 0.1 µm, improvements in materials of construction of process lines and tanks, and better container washing techniques.

In addition to customer quality control tests, many customers are moving towards purchasing specifications that include statements that the chemicals must be packaged in cleanrooms of Class 100, or in some cases Class 10.

3.8. ANALYTICAL CAPABILITIES

Along with higher chemical purity comes the concern regarding vendor analytical capabilities. Even today's specifications require quite sophisticated analytical instruments and highly trained chemists and chemical technicians to maintain quality control at the supplier level and for incoming inspection at the user level.

In addition to chemical purity, the other major technical issue concerns particulate levels in liquid chemicals. There is general agreement that particulates, regardless of their source, are a major factor in causing device failure in VLSI circuits. Much has already been done by suppliers to improve filtration, in pre-cleaning bottles and in improving the cleanliness of the environment in which packaging is done. Further improvements continue on an on-going basis.

Each user should be aware of the analytical capabilities and quality programs of its chemical supplier or proposed suppliers. This should be an important factor in choosing a vendor and it will sometimes determine the level of the required incoming inspection program that the user must install.

Vendor qualification procedures should focus heavily on the analytical capabilities of the vendor. Audits should be conducted periodically, and should focus on:

- Analytical instruments
  - Calibration
  - Maintenance
  - State-of-the-art

- Laboratory procedures and methods
  - Good record keeping and retained samples program
  - Standards preparation

- Accuracy and repeatability
  - Evaluate laboratory's performance

- Adequate staffing
  - Technician training
  - Personnel turnover

- Reporting structure
- — Does lab have final shipping authorization?

REFERENCES

Brandt, T., 1983: "The cleanroom of the future," *Microcontamination* **1** (3): pp. 27–29.

Cheung, S.D. and R.P. Roberge, 1986: "Measurement of particles in IC process equipment," Proc. Microcontamination Conference, Santa Clara, CA., November 18–21, pp. 130–146.

Davison, J. and J. Hoffman, 1987: "Ultrapure piranha for ULSI applications," Proc. of the First International Symposium on ULSI Science and Technology, Philadelphia, PA.

Dillenbeck, K., 1991: "Achieving chemical systems integration within wafer fabrication areas," *Microcontamination* **9** (4): 52–59.

Duffalo, J.M. and J.R. Monkowski, 1984: "Particulate contamination and device performance," *Solid State Technology* **27** (3): 109–114.

Fujimura, S. and H. Yano, 1986: "New device degradation mechanism: Heavy metal contamination from resists during plasma stripping," Electrochem. Soc. Proc., Sixth Symposium on Plasma Processing, Vol. 87–6, San Diego, CA.

Hashimoto, S., M. Kaya, and T. Ohmi, 1989: "Ultra-high-grade chemicals—Part II: Improving and maintaining electronics-grade chemical quality requires technological advances," *Microcontamination* **6** (7): 25.

Ito, H. and G. Wilson, 1982: Proc. SPIE Conference, "Photopolymers principles-processes & materials," Ellenville, N.Y., pp. 331.

Jain, S., S. Chatterjee, P-H. Lu, and E. Brown, "Advances in photoresist technology," *Microelectronic Manufacturing and Testing* **13** (10): 14–17.

Jones, A.H., J.G. Hoffman, A.W. Jones, and W. Yuan, 1987: "Improved wafer cleaning with ultra pure piranha," Proc. of Microcontamination Conference and Exposition, Santa Clara, CA, October 27–30, pp. 69–80.

Kern, W. and D.A. Puotinen, 1970: "Cleaning solutions based on hydrogen peroxide for use in silicon semiconductor technology," *RCA Review*, June, pp. 187–206.

Knollenberg, R.G., 1983: "In situ optical particle size measurements in liquid media," 2nd Annual Semiconductor Pure Water Conference, San Jose, CA, January 13–14, pp. 82.

Krygier, V., 1986: "High-purity water: rating of fine membrane filters used in the semiconductor industry," *Microcontamination* **4** (12): 20–26.

Long, M.L., 1984: "Photoresist particle control for VLSI microlithography," *Solid State Technology* **27** (3): 159–161.

Menon, V.B., L.D. Michaels, R.P. Donovan, and D.S. Ensor, 1989: "Effects of particulate size, composition, and medium on silicon wafer cleaning," *Solid State Technology* **32** (3): S7–S12.

Ohmori, T., 1989: "Ice scrubbing effective, flag raised over organic-vapor contamination," *Semiconductor International* **12** (11): 16.

Ong, E. and E.L. Hu, 1984: "Multilayer resists for fine line optical lithography," *Solid State Technology* **27** (6): 155–160.

Peters, L., 1992: "Stripping today's toughest resists," *Semiconductor International* **15** (2): 58–64.

Pruett, K.K., 1990: *Compass Corrosion Guide II: A Guide to Chemical Resistance of Metals and Engineering Plastics*, La Mesa, CA.

Rechen, H.C., 1985: "Microorganism and particulate control in microelectronics process water systems-pharmaceutical manufacturing technology," *Microcontamination* **3** (7): 22–29.

Tai, K.L., R.G. Vadimsky, and E. Ong, 1982: "Multilevel Ge-Se film based resist systems," SPIE Proc., Vol. 333, pp. 32–39.

Tolliver, D.L., 1984: "Contamination control: New dimensions in VLSI manufacturing," *Solid State Technology* **27** (3): 129–137.

Watts, M.P.C., 1984: "Electron beam resist systems—a critical review of recent developments," *Solid State Technology* **27** (2): 111–113.

Chapter 4

GASES

4.1. TECHNOLOGY ISSUES

Specialty and bulk gases are used in virtually every phase of semiconductor fabrication. There are four common methods of gas delivery from supplier to customer: pipeline, on-site plants, liquid tankers and cylinders. Hydrogen, nitrogen, argon, and oxygen are standard gases distributed to point-of-use by the first three methods whereas specialty gases such as silane and phosphine are supplied in volumes ranging from a single cylinder to a tube trailer.

Automation in gas cabinet and gas room design has become highly sophisticated, as companies such as IBM, AT&T, and Xerox have recently installed computer controlled systems. New trends include the remote monitoring of process purity and safety as well as on-site and remote alarm systems with full status accessibility for any gas cylinder in any cabinet (Boyd 1984). Several issues have been identified to assure proper gas system design as listed in Table 4.1 (Lorenz 1984).

Other requirements in assuring a clean gas system include:

- Scheduled Maintenance/Inspection

- Leak Tight Systems

Table 4.1. Gas control system issues.

Vents
   Vacuum assisted
   Arsine/phosphine or diborane piped to burn box
   Silane piped to burn box
   Reactive gases piped to scrubbers
   High and low pressure vents on purge panel
   Regulator bonnet vents
Alarms
   Excess pressure
   Excess flow
   Low process gas cylinder pressure
   Low purge gas cylinder pressure
   Blocked vent
Valves
   Vent check
   Diaphragm or bellows-sealed
   Ball valves not permitted

- Clean Gases
- Clean System Fabrication
- Correct Installation
- Personnel Training
- Automatic Purging
- 0.02 Micron Particulate Filtration
- Oxygen/Moisture, Oxides of Carbon, Nitrogen, and Sulfur Purification and Absorption

4.2. REQUIREMENTS

4.2.1. Purification Alternatives

Historical Perspective

In the 1980s, the improvement of Japan's market share was partially due to the superiority of their manufacturing methods. A portion of that competitive advantage was due to purer chemicals. While the Japanese were thinking and buying at ppb levels, the rest of the world was accepting ppm quality. This difference was noted by some of the leading-edge device manufacturers, both because of the Japanese and the necessity of purer chemicals to produce the newer devices at economical yields.

The chemical industry responded in a slow fashion with the introduction of improved trace element products. The slowness was attributable to the downturn of the semiconductor industry in 1984 and the poor financial return in this segment of the chemical industry not supporting the development needed for purer product technology.

The creation of Sematech can be looked as a turning point in the call for purer chemicals. Sematech's effort to bring together the users and producers to discuss and set forth goals to be achieved was successful. While the targets of Sematech are confidential, it did establish concrete goals in stages that were rational and understandable. The greatest benefit was a set of quantifiable specifications, not a subjective call for orders of magnitude improvement across all parameters.

Trends for Purity—Consistency

A new concern has emerged as the suppliers have improved the purity of the delivered product-consistency. The manufacturing of devices at adequate yields has become dependent on minimizing process variances. The current purity of gases has now reached a threshold where subsequent deterioration or improvement of purity, even in a single trace element, can upset the yield. The fact that improvement in purity can be viewed as a negative has been somewhat ironic to the suppliers.

From the supplier perspective, the need to control their manufacturing process so tightly is economically troubling. While adoption of statistical process control and statistical quality control programs has improved lot-to-lot consistency, the problem remains with what to do with product that does not meet a consistency specification. If only a few advanced technology customers adopt a consistency specification, the remaining customers can be sold the product. However, if a majority of the customers adopt a consistency specification, the problem of what to do with inconsistent product becomes a major economic challenge. The cost of monitoring a consistency specification in of itself is a major factor.

The high purity of gases available to semiconductor manufacturers has reached a level whereby it is not an immediate concern in the fab line provided they are consistent and meet the initial specifications at time of purchase.

In 1986, grain boundary stress corrosion creep cracking in 6351 alloy aluminum was discovered, dictating the conversion to 6061 alloy aluminum.

Other cylinder types included a 316L stainless steel from Union Carbide and a very low pressure (<240 psig) rolled tube (for use with WF_6).

Gas suppliers tailor packaging methods to a particular gas. Matheson uses polished-interior and coated cylinders for arsine, phosphine, silane, and ammonia. These are baked at 200° under a 10^{-15} Torr vacuum with a turbo-molecular pump and filled through a 0.1 µm PTFE filter through 316L stainless steel tubing.

Airco uses seven major steps to prepare a gas cylinder (Wechter 1990):

- Cylinder cleaning
  — Internal surfaces are washed using a sequence of high-purity halocarbons and DI water.
- Valve maintenance
  — Valves are cleaned in a Class 100 environment using an ultrasonic cavitation process.
- Particle characterization of cylinder and valve
  — Cleaned assemblies are analyzed for particulate contamination according to customer specifications.
- Cylinder evacuation
  — Each assembly is heated in a cylinder bakeout oven and evacuated to remove moisture and low-vapor- pressure organics. The process is microprocessor controlled. Residual gas analysis is performed by mass spectroscopy with a sensitivity of 1 ppb.
- Filling
  — Cylinders are filled in Class 10 laminar flow hoods. All fill manifolds are constructed of 316L electropolished stainless steel, and continuously purged with an inert gas when not in use. Microprocessor-controlled pneumatic manifold valves are driven by both gravimetric and manometric feedback loops to ensure accuracy and consistency of blends to 100 ppb.
- Gas analysis
  — Samples are measured to ppb levels for metallic and dopant impurities and ppm levels for other impurities.

A great deal of emphasis in high-purity levels of process gases deals with clean gas systems; gas handling equipment and gas cabinets. High quality (15 root mean square) surface finishes inside stainless steel tubing, fittings, pressure regulators, and valves are now required in order to prevent entrapment of particulates (Hardy *et al.* 1988). Use of freon PCA friction cleaning with filtered nitrogen and handling of components in a Class 100 cleanroom using double bagging can reduce particulates. New technology is being advanced with sophisticated processes of passivation, chemical polishing or electropolishing, and micropolishing. Socket welding of fittings has been replaced by automatic orbital tube welding under a blanket of filtered high-purity argon purge gas.

Carrier gases such as nitrogen, oxygen, and hydrogen can be further purified in-house with the use of in-line moisture traps, hydrocarbon traps, or purifying systems (Boyd and De Bord 1985). The point-of-use purifying systems provide a reliable and economical method of producing large volumes of ultra-high purity grade electronic gases from commercial grades. Rare gases such as helium and argon can also be purified with point-of-use systems from commercial grade to high-purity grade. This process is performed by: passing the gas over titanium granules at 700° C to remove nitrogen and oxygen; over a copper oxide catalyst at high temperature to remove hydrocarbons, hydrogen and carbon monoxide; and through a molecular sieve at room temperature to remove moisture and carbon dioxide.

4.2.2. Purity Trends

Most gas suppliers offer specialty products rated at five nines (99.999%) purity or better. Shown in Table 4.2 are product specifications for several specialty gases.

Most gas suppliers also offer specialty gases in various grades, depending on the application. Solkatronic supplies Arsine in three grades, as shown in Table 4.3. Their Megabit[R] grade, at 99.9999+% purity, has germanium impurities less than 50 ppb and silicon impurities less than 100 ppb, two important impurities in GaAs IC production for reduced background carrier concentrations.

Contamination in processing gases arising from moisture, oxygen, carbonaceous materials, metallics, and particles has adverse effects on ICs, resulting in undesirable changes in the formation of silicides and aluminum films, the growth of epitaxial films, and the overall quality of the devices.

Gas suppliers have succeeded in removing contamination to levels whereby the IC manufacturer can fabricate devices with consistently high yields. As devices become more complex and as feature sizes become smaller, additional improvements must be made. Sematech's goals for maximum impurity in bulk nitrogen, oxygen, argon, and hydrogen was <100 ppb in 1989, <10 ppb in 1990, and <1 ppb in 1993.

In meeting Sematech's needs for these gases, Linde uses cryogenic separation processes to produce gaseous and liquid nitrogen, oxygen, and argon, and liquid nitrogen. Purification methods for each are shown below (Hardwick et al. 1988):

Table 4.2. Purity specifications of specialty gases.

Silane ULSI SiH₄ (99.999%)			Arsine ULSI AsH₃ (99.999%)			Phosphine ULSI PH₃ (99.999%)		
$O_2 + Ar$	< 1	ppm	GeH_4	ND 50	ppb	O_2	< 1	ppm
N_2	< 2	ppm	SiH_4	ND 50	ppb	N_2	< 1	ppm
CH_4	< 2	ppm	C_2H_6	ND 50	ppb	CO	< 1	ppm
CO	< 0.1	ppm	C_2H_4	ND 50	ppb	CO_2	< 1	ppm
CO_2	< 1	ppm	S	ND 50	ppb	CH_4	< 1	ppm
H_2O	< 1	ppm	O_2	40	ppb	H_2O	< 2	ppm
Chlorosilanes	< 1	ppm	N_2	250	ppb	AsH_3	< 1	ppm
Typical resistivity	10,000	ohm–cm	CO_2	40	ppb			
			CH_4	ND 10	ppb			
			H_2O	350	ppb			
			PH_3	ND 50	ppb			

Hydrogen chloride ULSI HCl (99.999%)			Chlorine, USLI Purity Cl₂n (99.999%)			Ammonia, ULSI Purity NH₃ (99.999%)		
CO_2	< 6	ppm	CO_2	< 6	ppm	H_2O	< 3	ppm
N_2	< 1.5	ppm	CO	ND < .5	ppm	O_2+	< .5	ppm
O_2	< 1	ppm	CH_4	ND < 1	ppm	CO_2	< .4	ppm
CH_4	< .1	ppm	O_2	< 1	ppm	CH_4	ND < .1	ppm
H_2O	< 1.4	ppm	N_2	< 3	ppm	N_2	≤ 1	ppm
						CO	ND .1	ppm

Table 4.3. **Range of purity of CVD gases.**

Analysis	Megabit	MOCVD	Electronic
Arsine	99.9999+% min.	99.9998% min.	99.999% min.
Argon	N.D. < 0.1 ppm	N.D. < 1 ppm	–
Nitrogen	N.D. < 0.1 ppm	< 1 ppm	3 ppm
Oxygen	N.D. < 0.1 ppm	N.D. < 1 ppm	1 ppm
Carbon dioxide	N.D. < 0.1 ppm	N.D. < 1 ppm	1 ppm
Carbon monoxide	N.D. < 0.1 ppm	N.D. < 0.5 ppm	0.5 ppm
Methane	N.D. < 0.1 ppm	N.D. < 1 ppm	–
Ethane	N.D. < 0.1 ppm	N.D. < 1 ppm	–
Propane	N.D. < 0.1 ppm	N.D. < 1 ppm	–
THC	–	–	< 1ppm
Phosphine	N.D. < 0.1 ppm	< 2 ppm	2 ppm
Water	N.D. < 0.1 ppm	< 0.5 ppm	< 3 ppm
Germanium	< 50 ppb	–	–
Silicon	< 100 ppb	–	–

- Argon

    — Argon is recovered from a side column of an air-separation system in the purification of oxygen, and contains impurities other than oxygen and nitrogen in the ppb level. In a further refinement, hydrogen (in excess levels) is added to the gas stream to react with the oxygen over an oxidation-catalyst bed. The effluents at this point contain argon, nitrogen, excess hydrogen, and moisture, as well as small levels of methane (if introduced with the hydrogen) and carbon dioxide as a result of methane oxidation. A molecular sieve is then used to remove carbon dioxide and moisture. The stream is distilled cryogenically to reduce nitrogen levels to <1 ppm and to remove hydrogen. Any unoxidized methane is not removed.

- Nitrogen

    — Nitrogen contains ppm levels of hydrogen, helium, and neon in its gaseous form, while the liquid form has essentially no levels of these impurities. Both forms, however, contain argon and <1 ppm oxygen and carbon monoxide. When distilled, other contaminants such as carbon dioxide, moisture, oxides of nitrogen, hydrocarbons, krypton, and xenon are less volatile than oxygen and are reduced to sub-ppb levels.

- Hydrogen
  — Only helium and neon have a solubility >100 ppb in liquid hydrogen. The stream is purified to this level using several cryogenic adsorption traps in series.
- Oxygen
  — Absorption traps are used to remove hydrocarbons, carbon dioxide, and moisture prior to a distillation process that reduces hydrocarbons to 15 ppm, carbon dioxide to <100 ppb, and krypton and xenon to 5-10 ppm levels. Additional cryogenic distillation steps lower the hydrocarbon and carbon dioxide and levels of argon, krypton, and xenon. Argon is typically removed to <2000 ppm.

Point-of-use purification can be utilized to further improve the purity of gases, removing contaminants from the above sources as well as from the cylinder. The following are a compilation of purification methodologies for various gases:

- Argon and Nitrogen
  — Oxygen scavengers can be employed using a reduced copper catalyst for argon and nitrogen. This method can maintain critical impurities below 0.1 ppm. However, the scavenger must be properly activated by heating in a large, complex system that requires high levels of maintenance.
  — Argon can be purified using titanium (and other metal alloy) getters. This method can maintain critical impurities below 1 ppm. However, a flow capacity of <250 sccm is a limitation.
  — Activated molecular sieves can be used to reduce moisture and carbon dioxide. This method can maintain critical impurities below 0.1 ppm. However, the sieve must be properly activated by heating in a large, complex system that requires high levels of maintenance.
  — A copper oxide catalyst can be used to oxidize carbon monoxide and hydrogen to carbon dioxide and water, which are then removed by a molecular sieve.
  — Highly porous organometallic polymers react instantly with oxygen, water vapor, chlorosilanes, chlorofluorocarbons, dopants, and, to a lesser degree carbon monoxide. These resins contain reactive sites of carbanions and hydride ions that are both highly reducing and basic. This method can maintain critical impurities below 10 ppb for oxygen and water vapor.
- Hydrogen
  — A platinum catalyst can be used to convert oxygen to water that is then removed by a molecular sieve. This method requires a regenera-

tive system. Another limitation is the problem of assuring that the platinum catalyst is not exposed to hydrogen mixtures containing more than one percent oxygen.
— A heated palladium alloy tube is used to purify hydrogen to impurity levels of <10 ppb. Hydrogen diffuses through the walls of the tube heated at temperatures of 450°C. Three limitations include safety factors with hydrogen above 459°C, microcracking of the tube due to a phase change in the presence of hydrogen, and high pressures of 250 psig to achieve maximum flow rates.
— Highly porous organometallic polymers discussed above. This method can maintain critical impurities below 10 ppb for oxygen and water vapor.

- Silane and Ammonia
  — Molecular sieves and activated carbon have been explored on the research level for silane. Ammonia is absorbed on molecular sieves like water and cannot be removed.
  — Potassium hydroxide and distilled sodium metal have been used to purify ammonia in the laboratory.
  — Pyrolysis can be used to crack arsine and phosphine in silane.
  — Highly porous organometallic polymers discussed above.

Gas suppliers have begun addressing gas delivery in terms of a system-wide concept, whereby they have invested heavily in:

- Advanced manufacturing—Companies have started a concerted effort on the utilization of the highest purity raw materials available, and further purifying them. Detail has been placed on packaging in Class 100 cleanrooms and Class 10 hoods. Statistical process control methods have been incorporated to assure purity and reduce lot-to-lot variability.

- Advanced analytical equipment—Suppliers now use an array of analytical techniques to assay their products, such as gas chromatography, mass spectroscopy and infrared spectroscopy. New equipment includes ultrasonic detectors with improved resistance to gas mixtures and highly sensitive helium ionization detectors.

- Advanced packaging—Specially treated cylinders for ultra-high purity grades of gases is becoming the norm. Gas suppliers use a proprietary process for coating or polishing the inside of a gas cylinder. These are then baked at temperatures of 200° under an ultra-high vacuum. As the gas is injected through stainless steel piping and bakeable block diaphragm valves, it is filtered through 0.1 µm PTFE filters.

These advanced techniques are responsible for supplying the fab with the cleanest gas currently available. What becomes a necessity for fab engi-

70   GASES

neers is to **keep the same level of quality of gas as it is received**. This suggests that:

- Gas suppliers and fab engineers cooperate in defining quality. Engineers must provide delivery systems to point of use that maintains the high level of purity in the gas cylinder.

- Gas suppliers and equipment vendors cooperate in maintaining high purity to insertion in the reaction chamber; i.e., equipment plumbing must incorporate the same high levels of control as delivery lines including point of use filters.

- Gas and equipment suppliers and users must work in a concerted effort to:
  — Understand contamination principles
  — Understand thin film deposition and etching chemistry to develop new chemical compounds or gas mixtures
  — Understand gas kinetics in an attempt to lower processing temperatures, gas phase contamination, parasitic reactions, intermediates, and precursors

4.2.3. Particulate Considerations

The current SEMI particle specifications for VLSI gases is <20 particles per cubic ft ≥ 0.2 µm. However, gases at point-of-use should have typical particle levels of <10 particles per cubic ft ≥ 0.1 µm. In fact, technology is available to achieve <1.0 particle per cubic ft ≥ 0.1 µm. This can be achieved by a concerted effort using:

- Aerosol filtration technology
- Refined component selection and installation
- Particle counting methods

Particle Monitoring

Optical particle counters or continuous-flow condensation nucleus counters can be used to monitor particulates in gases. The optical particle counter is the most widely used method, both in gas supply systems and in cleanrooms. A light source illuminates a small viewing volume and a photodetector measures the scattered light from the individual particles as they pass through this illuminated volume. The detection limits are 0.1 to 40 µm using a modified standard counter. The continuous-flow condensation nucleus counter oper-

GASES

ates by passing the gas through a heated alcohol vapor and a refrigerated condensor. Condensed alcohol droplets emerge and are counted by an optical system similar to that used in the method above. These droplets are able to condense on particles as small as 0.1 µm. Condensation nucleus counters do not have size-discriminating capabilities and thus count all particles larger than 0.01 µm.

Particle detection on the wafer surface is an additional method of monitoring particulates after the contaminant has condensed (Gise 1983). A focused He–Ne laser beam scans the substrate surface in a raster pattern. On a contaminated surface, the laser beam is scattered and collected by a photomultiplier tube with a particle-size sensitivity of 0.5 µm.

The handling of these gases poses both safety and health dangers to personnel. These gases are classified in Table 4.4.

Filtration Methods

Particulate removal is one of the most strategic issues facing both suppliers and users of gases in semiconductor fabrication. Carrier gases are usually stored in bulk and piped into processing areas. They pass through piping, compressors and connections, and can contain particles from solder debris, rust, metal shavings, moisture and filters. Semiconductor-grade bottled gases are lower in particulates, but impurities can be generated from the movement of mechanical parts such as regulators, valves, and flow controllers. The incorporation of gas-derived particles often leads to crystallographic defects resulting in stacking faults, dislocations and metal precipitates.

Table 4.4. Potential hazards of processing gases.

Oxidants
    Air, Cl_2, N_2O, O_2
Flammables
    AsH_3, B_2H_6, GeH_4, H_2, PH_3, $SiHCl_3$, $SiHCl_3$, SiH_4
Explosives
    CH_4, C_2H_2, CO, H_2, H_2S
Corrosives
    BCl_3, BF_3, Cl_2, HCl, H_2S, NH_3, PCl_3, PCl_5, $SiCl_4$, $SnCl_4$, WF_6
Toxics
    AsH_3, BCl_3, BF_3, B_2H_6, Cl_2, CCl_4, $CHClF_2$, CO, H_2S, HCl, NH_3, PCl_3, PH_3, $SiCl_4$, SiH_4, SiH_2Cl_2, WF_6
Malodorous
    BF_3, Cl_2, HCl, H_2S, NH_3, N_2O
Suffocating
    Ar, CO, N_2

Filter types include stacked-disc PVDF membrane, pleated microfilament membrane, and fiberglass filters. Stacked-disc PVDF membrane filters are most often used and demonstrate 0.05-μm filtration with extremely high efficiency with both toxic and non-toxic gases (Zuck 1989). Sealed stacked-disc PVDF membrane filters are recommended for reactive gases such as HCl, phosphine and dichlorosilane. They should be placed in series in the gas line and changed at regular schedules, ranging from 3 to 12 months, depending on gas cleanliness, gas flow, and processing step (Accomazzo *et al.* 1984).

REFERENCES

Accomazzo, M.A., K.L. Rubow, and B.Y.H. Liu, 1984: "Ultrahigh efficiency membrane filters for semiconductor process gases," *Solid State Technology* **27** (3): 141–146.

Boyd, H., 1984: "Non-contaminating gas containment, control, and delivery systems for VLSI-class wafer fabrication," *Microelectronic Manufacturing and Testing* **7**, March, pp. 29–34.

Boyd, H. and D. DeBord, 1985: "Process gas analysis for VLSI wafer fabrication," *Microelectronic Manufacturing and Testing* **8** (5): 1–13.

Gise, P., 1983: "Principles of laser scanning for defect and contamination detection in microfabrication," *Solid State Technology* **26** (11): 163–165.

Hardwick, S., R.G. Lorenz, and D.K. Webster, 1988: "Insuring gas purity at point-of-use," *Solid State Technology* **31** (10): 93–97.

Hardy, T.K., D.D. Christman, and R.H. Shay, 1988: "Measurement and control of particle contamination in high purity cylinder gases," *Solid State Technology* **31** (10): 83.

Lorenz, A.K., 1984: "Maintenance of gas purity in the fab area," Microcircuit Pure Materials Conference, San Jose, CA, August 2–3.

Wechter, S., J. Jordon, and T. Seidler, 1990: " Handling ULSI-grade electronic gas cylinders," *Microelectronic Manufacturing and Testing* **13** (5): 71–72.

Zuck, 1989: "Particle control in the construction of a 1 Mbit DRAM gas distribution system," *Solid State Technology* **32** (11): 131–135.

Chapter 5

LITHOGRAPHY

5.1. OPTICAL SYSTEMS

Optical steppers for wafer imaging have remained the dominant force in the IC industry despite claims by electron-beam, X-ray, and focused ion beam equipment manufacturers that high resolution in needed for VLSI devices. Unit shipments can attest to the continuing demand for optical stepper technology: there is an installed base of 5,700 steppers, and sales in 1991 added another 700 units. Contrast this to non-optical methods, with E-beam used for low-volume specialty products and X-ray just entering the marketplace for R&D applications.

One of the main reasons for the continual acceptance of this methodology has been the advances in the exposure wavelength of the ultraviolet radiation (UV) available with these systems. Whereas exposure limits of 0.6 µm are possible in production using near-UV (330–450 nm) radiation, extending the radiation to the deep-UV (DUV) range (200–260 nm) permits linewidth resolution of excimer laser technology to extend optical capabilities well below 0.5 µm.

A significant factor in the continual success of optical microlithography is its maturity. There is a strong database in the semiconductor industry regarding optical lithography. A large installed base of optical systems, coupled with technological advances, lower purchasing price, more economical operation, and the considerable development required for masking and resist techniques for E-beam and X-ray techniques add to the success of the optical steppers.

Advances in optical stepper systems and in resist materials, have been another factor in addition to DUV exposure in maintaining the dominance of optical lithography (McCoy, 1989). These include:

- Higher throughput
- Higher wattage lamps
- New lens design
- Computer controlled optical adjustments
- Small footprint
- Improved automatic field-by-field alignment
- Mix-and-match alignment strategies
- Wide exposure latitude resists
- Multi-level resist processes

As production quantities of 4 Mbit DRAMs began in 1989, lithography requirements included:

- Minimum Feature—0.8 µm
- Total Overlay—±0.25
- Critical Linewidth Control—±0.10
- Minimum Field Diameter for 2 Die—2.0 cm

Described below are the technological issues and trends in optical methods of microlithography that are underway in order to meet these objectives. Also included are system description and analysis and photoresist trends.

5.1.1. Proximity/Contact Aligners

Proximity and contact alignment are the oldest methods of imaging (Markle 1974) (King 1979) used in semiconductor fabrication. Although the usage of these methods has decreased in the past few years due to fears in the industry that these systems are performance limited, there is continued use in advanced processes because of simplicity of use, low cost, and recent equipment advances that achieve less than 1 µm resolution and less than 0.5 µm registration using DUV radiation.

LITHOGRAPHY

State-of-the-art contact aligners have a resolution in the submicron range. A commercial system from Karl Suss is available with a resolution as low as 0.2 µm using an excimer laser source. However, yields are low compared to other submicron methods such as E-beam. The limitation of proximity aligners is approximately 2 µm using DUV radiation. At this wavelength, there are less deleterious diffraction effects caused by the shadowing of the parallel light beam by the pattern of the mask. However, even with these advancements in linewidth resolution, several problems arise when using DUV radiation:

- Photoresist sensitivity is reduced
- Expensive quartz masks are required
- Registration decreases on large wafers

Other disadvantages of these methods, particularly with contact aligners are:

- Mask damage
- Inconsistent masks
- Wafer bowing
- Poor realignment capabilities
- No scale adjustment

Although there are limitations to the proximity alignment technique, advantages include:

- Less wafer or mask damage compared to contact exposure
- High throughput
- High uptime
- No imaging optics
- Low cost (approx. $30,000)
- Compatible with stepper aligners in a mix-and-match arrangement

The use of proximity/contact aligners on IC production lines will be phased out over the next several years and their use limited to startups and R&D facilities who are working on 3-inch wafers. Non-Si IC devices will continue to be processed with contact/proximity aligners including hybrids, and materials such as garnet, sapphire, and graphite in which 3-inch diameter wafers are the largest available commercially. GaAs devices in the high-frequency range with few features will continue to use contact aligners, where particles generated from the mask will not present a problem.

5.1.2. Scanning Projection Aligners

A scanning projection aligner utilizes a scanning rather than a full-field exposure method to project the image of the mask features onto a wafer in order to minimize distortions in the image (Burggraaf 1989). The 1X optical system uses a spherical, reflecting mirror surface to image the mask onto the wafer in which both move together by means of a continuous scanning mechanism.

As with proximity/contact aligners, resolution depends on the wavelength of the illumination source. At 400 nm, minimum resolution is approximately 1.25 µm on commercial systems. At 260 nm (DUV), resolution is slightly less than 1 µm and with an overlay accuracy of 0.25 µm (Resor and Tobey 1979). The high resolution capabilities of this method makes it suitable for a mix-and-match arrangement with steppers. Another advantage is high throughput—up to 100 6-inch wafers per hour.

Projection aligners were used for all process levels in the production of 64 Kbit and 256 Kbit DRAMs with excellent and reproducible resolution and overlay. However, for denser circuits, aligners are being used for non-critical layers where resolution and registration are not crucial.

For ASICs, the prohibitively high cost of 6- and 7-inch mask sets for scanners will impact sales, so that steppers will be used for feature sizes well within the resolution of scanners. Nevertheless, ASICs have generally larger die sizes, and aligners are a more effective tool for the fabrication of much larger dies than a stepper, which is ideal for memory devices.

Sales of scanning projection aligners will continue to be made despite the emphasis on steppers because of recent technological advances including DUV and excimer laser sources, new resists, new lenses, contrast enhancement coatings, 6-inch wafer capability, and mix-and-match compatibility with steppers in non-critical mask applications.

Step-and-scan is a radically different approach to lithography (Buckley and Karatzas 1989), combining advantages of both scanners (large field size) and steppers (high NA reduction optics and field-by-field alignment, focusing, and leveling (Buckley and Karatzas 1989). Perkin–Elmer's (SVG Lithography) Micrascan I utilizes a 2400W mercury–xenon short-arc lamp as a deep UV (DUV) source, enabling 0.5-µm resolution, with a machine-to-machine overlay accuracy of 0.5 µm, and a depth of focus of ±0.75 µm. The exposure illumination is over a relatively broad 240 nm to 255 nm wavelength bandwidth. The field size is extremely large—20 mm by 32.5 mm—making throughputs of 35 to 50 8-inch wafers per hour feasible.

LITHOGRAPHY

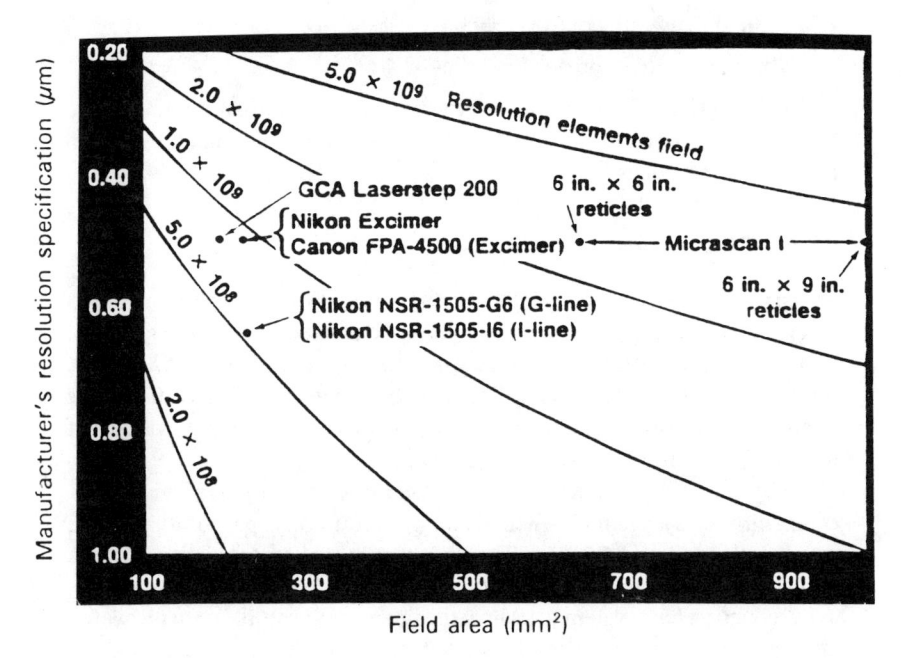

Figure 5.1. Resolution versus field area.

The system scans a field on the wafer past the fixed arcuate image field of a catadioptric projection optics configuration with a 4:1 reduction ratio. The reticle stage and wafer stage are both in motion, but the reticle moves four times as fast as the wafer.

The broad bandwidth of the optics, coupled with the broadband illumination, circumvents the standing-wave problem that is characteristic of monochromatic systems such as excimer lasers. The wavelength is monitored with a spectrometer, which feeds information to a wavelength-control system to keep the wavelength within 0.05 angstroms.

The large field size of the Micrascan system corresponds to 2.6×10^9 resolution elements and requires 6 in. $\times$ 6 in. reticles. As shown in Figure 5.1, larger reticles will enable the field size to be increased to 20 mm $\times$ 50 mm, corresponding to 4.0×10^9 resolution elements.

Production quantities of 16 Mbit DRAMs began in 1991. Lithography requirements include:

- Minimum feature—0.5 µm

- Total overlay—±0.15

- Critical linewidth control—±0.05
- Minimum field diameter for 2 die—2.4 cm

There will a shift from stepper to step-and-scan utilization to achieve these goals. Equipment was purchased in 1989 for developmental work on these 16 Mbit devices.

5.1.3. Step-and-repeat Aligners

Unlike proximity/contact or projection aligners that use masks that are only a 1:1 (1X) representation of all IC images to be printed on the wafer, optical steppers have an additional capability of utilizing masks that are 5X or 10X times the size of the image (as well as reductions between 1X and 10X). This permits higher resolution images to be printed while reducing the effects of mask flaws or particles on the imaged wafer.

Step-and-repeat aligners, illustrated in Figure 5.2, use a refractive projection exposure system in which the image resolution is proportional to the wavelength and lens aperature. The typical exposure system incorporates a light source, a collimating lens system to focus the illumination on a reticle, and a reduction lens system to project the image of the reticle onto the wafer. After each exposure, the wafer is stepped relative to the mask, realigned, refocused, and exposed (Mayer 1980).

There has been a shift in thinking among users that reduction steppers are needed in submicron lithography. This is not the case (Voisin 1990), and 1X steppers have process characteristics comparable to reduction stepper with the ability to process submicron features. A cost analysis of 1X and reduction steppers shows that the yearly costs and investment costs of the reduction steppers can be twice that of 1X steppers. This cost analysis, performed by Ultratech Stepper, is given in Table 5.1.

As lithographic requirements for state-of-the-art devices become more stringent, some of the fundamental parameters in lithography take on added importance; linewidth and depth of field, determined by the equations (Lin, 1988):

$$L = k_1 \lambda/NA \qquad (1)$$

$$\Delta Z = \pm k_2 \lambda/(NA)^2 \qquad (2)$$

where L is the feature size to be printed, λ is the wavelength, NA is the numerical aperature of the lens, k_1 is a constant dependent on the quality and type of optical system and on processing conditions, Z is the depth of focus, and k_2 is another constant that has a typical value of 0.7.

On 5X to 10X reduction lens systems, the resolution depends exclusively on the numerical aperature (NA), not on the reduction ratio (Markle 1986).

LITHOGRAPHY

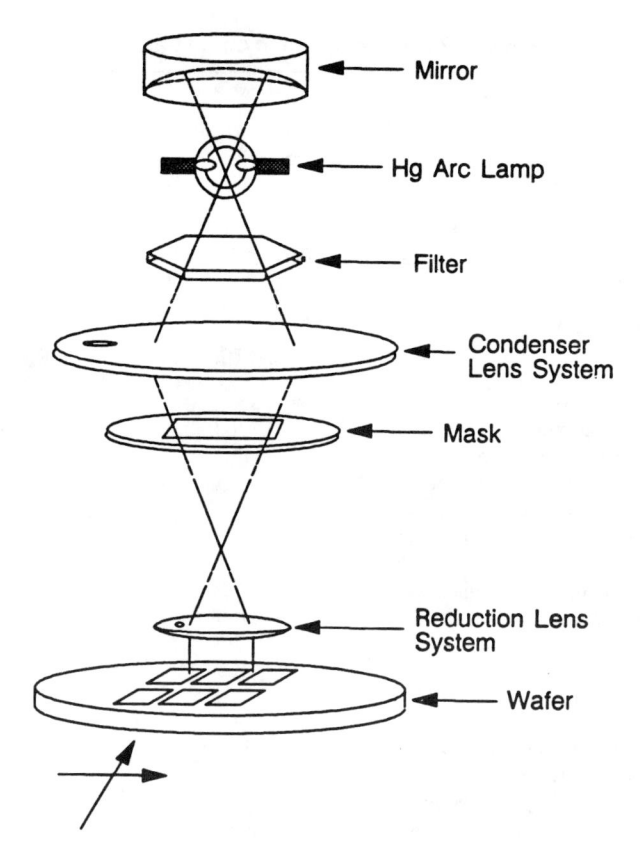

Figure 5.2. Schematic of step-and-repeat aligner.

However, although a higher reduction ratio permits a higher NA, a decreased depth of field can result in a need for critical focusing and leveling of the wafer. The depth of field is inversely proportional to the square of NA, as shown in Equation (2).

A significant advantage of the step-and-repeat aligners is that, by restricting the area of the image, an improvement in resolution is capable even on non-flat or contoured wafers (Wilson 1986). However, increased density of devices results in increased die size, and this is no longer true. Device manufacturers are resorting to planarization techniques that add complexity and cost to the process in order to attain flatter surfaces. Silicon wafer manufacturers are also burdened with the need for producing flatter wafers

LITHOGRAPHY

Table 5.1. Photolithographic cost analysis.

Operating Parameters:

Wafer size	6"
Average chip size	12 × 6 mm
Process layers	13
Production hours/week	100
New device layers/month	5
Planned wafer output/month	10,000
Operating environment	Memory

Cost Analysis:

	Model 1100/0.8 μm	Reduction
List purchase price	$850,000	$1,400,000
Depreciation costs/year (5 year straight line)	$170,000	$280,000
Instantaneous throughput (WPH) (<90% wafer coverage with site-by-site align, focus, level)	35	16
Throughput with EGA		29
Unscheduled maintenance	7%	7%
Job setup	10%	11%
Preventative maintenance	2%	5%
Waiting on wafers	10%	10%
Operator breaks	5%	5%
Total effective utilization	66%	62%
Effective throughput (WPH)	23.10	17.98
Wafers out/week	178	138
Number of steppers required	13	17
Total output capacity/month	10,010	10,189
Total Investment	**$11,050,000**	**$23,800,000**

Cost of Ownership:

Depreciation/wafer out	$19.09	$40.08
Consumables/month/system	$1,000.00	$4,000.00
Consumables/wafer	$1.30	$6.80
Maintenance/month/system	$2,000.00	$4,000.00
Maintenance/wafer	$2.60	$6.80
Direct labor/wafer	$17.69	$26.40
Engineering overhead/wafer	$8.84	$13.20
Facilities depreciation/month/system	$708.33	$2,125.00
Facilities depreciation/wafer	$0.92	$3.61
Resist cost/wafer	$6.00	$8.00
Reticle costs/month	$6,000.00	$9,000.00
Stepper cost/wafer out	$57.05	$105.79
Exposure Costs/Year	**$6,845,430**	**$12,694,558**

and minimize problems that could cause wafer warpage. As smaller features are required, increasing NA will result in smaller processing tolerances.

G-line (436 nm) 5X reduction steppers are currently used to define 0.8 μm features on 4 Mbit DRAMs. With numerical apertures above 0.40 and using Novolac resists, level-to-level overlay accuracy is in the range of from 0.25–0.35 μm. This is satisfactory for devices such as the 4 Mbit DRAM. However, an important issue is how far G-line can be pushed in order to obtain adequate processing tolerances and thus manufacturability.

The value of k_1 is an important factor, as this value is directly proportional to feature size. Reducing k_1 has been achieved by planarized surfaces, thin resist, and complex systems. However, a value of $k_1 = 0.8$ is typical in a development environment while 0.9 is normal in production. Therefore, according to Equation (1), *for a resolution of 0.8* μm, different values of k_1 equate to different NAs as shown:

k_1	NA
0.7	0.38
0.8	0.44
0.9	0.49

Thus, lenses with NAs of 0.44 can readily produce linewidths of 0.8 μm and a reasonable depth of focus (from Equation (2) with $k_2 = 0.7$) of ±0.79 μm.

At 0.5-μm resolution, these values equate to

k_1	NA
0.7	0.61
0.8	0.70
0.9	0.78

and the depth of field of ±0.32 μm. This is unreasonable. Current processing capabilities place a limit on a depth of focus of ±0.40 μm. Improved die-by-die leveling systems and aggressive development of new photoresist formulas are improving G-line's depth of focus however. Nevertheless, 0.5-μm capability will not be reachable in production.

G-line equipment will reach its limit at 4 Mbit DRAM and equivalent devices for critical mask levels. What the replacement for G-line will be is the most important issue facing 16Mbit production at 0.5 μm minimum feature sizes.

A recent innovation in steppers is the introduction of I-line lenses that operate at 365 nm. Although illumination is only in the near UV, better resolution is obtained. The I-line at 365 nm is 50% brighter than the G-line and has a narrower spectral linewidth. Since the sensitivity of photoresist is dependent on the intensity and spectral linewidth of the illumination, reduced standing-wave effects and improved step coverage is attained.

Current I-line lenses have a practical resolution near 0.75 μm. New lenses under development are predicted to have a practical resolution between 0.5 and 0.6 μm. Resorting to Equations (1) and (2), at *0.5-μm resolution*, these values equate to:

k_1	NA
0.7	0.51
0.8	0.58
0.9	0.66

The depth of focus is calculated to be ±0.38 μm.

There is increased concern over the types of glasses available with an NA of 0.66 that are I-line transparent, the ability to make the necessary aberrational corrections, and the effects of solarization and fluorescence (Wilson, 1986).

However, a combination of phase shifting masks, still in early stages of development, 0.48 NA lenses, and high contrast resists will make it possible to extend I-line equipment for 16 Mbit DRAMs. I-line equipment will be utilized for 16 Mbit DRAMs mask levels with 0.5 μm minimum feature sizes and possibly to 64 Mbits.

With phase shifting, masks have two different transparent areas, as opposed to standard masks with one opaque and one transparent. One of the transparent areas is thicker than the other. Light from the stepper source moves through the thicker portion more slowly than through the thinner region. Adjusting the thickness of the regions enables the light going through the thicker region to emerge 180° out of phase with the light through the thinner region. Because light bends as it travels through the mask, light from the two regions slightly overlap where they hit the wafer. Since they are out of phase, peaks of one beam are canceled by the valleys of another resulting in a very narrow dark spot. Unfortunately, a problem with phase shifting is the difficulty in manufacturing and testing masks. Faster RISC-based workstations are speeding up the mask pattern data processing.

High resolution is important only when all factors are considered including overlay accuracy, machine-to-machine accuracy, and registration. Factors responsible for low registration or accuracy are:

- Stepping stage quality
- Die placement error
- X-Y alignment
- Reduction error

LITHOGRAPHY

- Lens distortion
- Image rotation
- Alignment method
- Stage runout
- Focus error
- Thermal runout

Field size is an important criteria for high throughputs. Since the reticle of a 5X or 10X reduction system is 5 or 10 times the device image size, the size of the field depends on the reticle area. A 5X system has a typical field size of 15 mm × 15 mm (Canon's VL-20 and VL-21 have a field size of 20 mm × 20 mm) requiring a reticle area of 70 mm × 70 mm that is accommodated on a standard 5 inch reticle substrate. A 10X system, on the same size reticle substrate, has a maximum field area of 10 mm × 10 mm. This requires additional steps to cover the same area of a wafer and lower throughput.

The 4 Mbit DRAM is being produced by G-line 5X steppers with a 225 mm^2 field with two chips (100 mm^2 chip area each) per field. 16Mbit DRAMs have a chip area of 200 mm^2, making the Canon lenses suitable for two chips per field. For the 64 Mbit DRAM with a chip area of 400 mm^2, a throughput of only one chip per field will be achieved (McCoy 1989).

As feature sizes are reduced, there will not be a significant improvement in field size. As a result, it will be necessary to develop a field stitching strategy that does not require relaxed design tolerances at boundaries. It will also necessitate very short changeover times for reticles to compete with the very large field provided by step-and-scan systems (see Figure 5.1).

One area of concern with today's steppers is resolution inaccuracies due to lens distortion, magnification error, temperature, and atmospheric pressure. A typical stepper has an optical system that may contain 15 glass elements made of 6 different glass types each with a different diffractive index and dispersion. To control aberrations, the position of each element must be carefully adjusted (Ittner 1977). Small changes in the index of the glass that can occur from melt to melt must be measured and compensated during fabrication (Markle 1986). The index of the glass is also a function of temperature. A high throughput stepper requires that a high optical power level be passed through the lens. Any absorption of this power will change the temperature of the lens and upset the optical correction of the element.

For high-density devices, steppers must be totally characterized in order to control all correctable errors attributable to optical column alignment errors. This analysis must include not only misalignment of the reticle relative to the reduction lens, but to intrafield errors within the column itself. Intrafield matching error, defined as the relative placement pattern error caused by the difference in intrafield distortion between two steppers, contributes to stepper registration error to the same degree (0.05 µm) as system alignment error. The components of intrafield distortion are:

- Lens distortion error
- Magnification error
- Translation error
- Trapezoid error
- Residual error

Magnification and translation error can be minimized by optimization of alignment mark location and the use of a stepper magnification controller. Magnification error can also be corrected for by controlling the atmospheric pressure around the lens.

Laser-based lithography is a new technology that will extend the lifetime of optical lithography (McCleary 1988). Conventional coherent lasers are not used for lithography as they result in speckle; interference patterns that arise from varying optical path lengths that make it impossible to produce large-field light uniformity to the submicron range. Excimer lasers are different and deplete their pumped states faster than the time required for multiple reflections between the laser mirrors. Thus, the coherence length is negligible and these lasers operate more like a light bulb with a highly collimated light.

Again referring to Equations (1) and (2), at 248 nm and *0.5-µm resolution*, these values equate to:

k_1	NA
0.7	0.35
0.8	0.40
0.9	0.47

The depth of focus is calculated to be ±0.54 µm.

Excimer lasers have the advantages of:

- Higher intensity and thus shorter exposure times

- Flash times of microseconds eliminates the need for maintaining the stage stationary for several hundred milliseconds or wait for the alignment system to stabilize

- Conventional chrome-on-glass mask-making can be used as opposed to X-ray lithography requiring difficult-to-make boron nitride masks. However, if an ArF excimer laser is used, quartz masks rather than soda lime or borosilicate glass are required.

- A very high contrast (gamma > 10) so that new inorganic resist and resist processes can be used that depart from the narrow confines of traditional organic resists. When a 100 nm-thick metastable Al/O film is exposed to a single 20 ns ArF laser pulse, solid transformation takes place and the resultant profile has 0.5 µm feature sizes (Yokoyama *et al.* 1985).

Resist exposure with a XeCl excimer laser has been found to obey the law of reciprocity, i.e., the product of the light intensity and the required exposure time is intensity dependent (Rice and Jain 1984). Thus, the higher flux available for excimer lasers can be used to reduce the required exposure times. Yokoyama *et al.* (1985) have reported this feature with both conventional and excimer system (e.g., Al/O film) resists. These systems may undergo many orders of magnitude sensitivity enhancements as the dose rate is increased by increasing the incident pulse energy.

Several types of lasers operating in the deep UV region are used in lithography:

- Xenon Fluoride (XeF)—351 nm
- Xenon Chloride (XeCl)—308 nm
- Krypton Fluoride (KrF)—248 nm
- Argon Fluoride (ArF)—193 nm
- Fluorine (F_2)—157 nm

An additional problem associated with laser-based steppers is the transmission characteristics of the lenses. Two combinations of lens and laser sources have been produced:

- Quartz lenses and spectrally line-narrowed KrF laser
- Quartz and calcium fluoride lenses and broadband KrF

A quartz lens system is compatible with existing technology in G- and I-line lithography. However, problems in lens fabrication discussed above are compounded when lasers replace lamps. The purity of the lens material is tantamount to successful implementation of this technology. Impurities and atomic defects cause solarization such as color centers in the quartz. Also, the lens must be larger in diameter and requires a highly homogeneous refractive index of >10^{-6} when used with the finer resolution ArF excimer laser.

The quartz and calcium fluoride system, developed by Toshiba, is a 10:1 achromatic projection lens of 0.37 NA for the 248 nm wavelength. These chromic-aberration-corrected lenses can be made from other low-dispersion materials such as magnesium fluoride or lithium fluoride. Nevertheless, these lenses are difficult to manufacture with high quality and large diameters.

Excimer-laser lithography also has its set of problems (Jewell *et al.* 1987):

- The present lifetime of 10^9 shots is too short.
- A higher NA is required at 248 nm for finer resolution, dictating the need for a narrower bandwidth than the present 0.002 nm.
- Inner components and mirrors become contaminated and degraded by the reactive gas plasma.
- An absorption spectrum at 190 to 320 nm occurs with ArF due to the formation of point defects in the silica.
- Birefringence develops in unexposed areas due to macroscopic volume changes induced by the laser.

Current developments in the use of excimer-based lithography in the U.S. are:

- ASM Lithography introduced the PAS 5000/70 stepper that incorporates a KrF source and a 5X lens with a NA of 0.42, giving a field size of 21.2 mm. The $2.5 million system has already been delivered to AMD (Sunnyvale, CA) and IMEC (Leuven, Belgium).
- Researchers at AT&T Bell Laboratories have modified a commercial GCA 4800 stepper with a DUV KrF laser projection system that has defined 0.35-µm lines and spaces. The wavelength of the light is 248 nm, compared to 365 nm for I-line steppers. The 5:1 projection lens has a variable numerical aperature of 0.20 to 0.38 and a field size of 14.5 mm to 20 mm.
- Canon has developed the FPA-4500, capable of 0.5-µm resolution. The system is under evaluation, and has recently been sold to TI and to R&D facilities.
- GCA's AutoStep 200 is a combination UV/I-line stepper; a 5X reduction stepper equipped with a 2145 Tropel lens (21 mm diameter, 0.45 NA). Based on results with the Sematech program, GCA introduced the XLS 7000 Series for advanced memory and ASIC wafer fabrication. It features a wide field reduction lens for operation at either I-line or DUV. Practical resolution is achieved at 0.5- and 0.35-microns levels, respectively.
- Nikon developed the NSR-1505EX stepper with resolution better than 0.5 µm using a KrF excimer laser operating at 258 nm. The stepper sells for $2.3 million. Two systems are in beta sites at IBM (East Fishkill, NY) A wider-field NSR-2005EX has evolved.

LITHOGRAPHY

- Ultratech Stepper is developing a DUV stepper, but the light source is a filtered mercury arc lamp rather than an excimer laser.

Ateq Corp. (now the Laser Beam Microlithography division of Etec) adopted its Core 2500 laser-based reticle writing system to write directly on wafers. The Waferwrite 6000, priced at $2.7 million, is directed toward prototype and process development applications. The system is capable of writing patterns as small as 0.5 μm on 8-inch wafers with an overlay accuracy of 0.125 μm. The Waferwriter 6000 has a 0.60 NA aperature and a 20-to-1 lens combined with a 363.8 nm argon ion laser.

5.1.4. Mix-and-Match

Microlithography processes are highly capital intensive, with optical systems costing several million dollars. In order to delay capital outlay for new equipment and maintain low operational expenses, the asset life of older aligners is being stretched with the concept of mix-and-match lithography. This has resulted in fewer scanners purchased for fab lines.

Furthermore, maintenance costs are lower, due to the established base of existing equipment. However, mix-and-match often stirs up controversy in the fab area from maintenance personnel who must become proficient with two types of complicated equipment.

Several mix-and-match alternatives described below exist in semiconductor fab lines:

Projection Aligner-Stepper. A typical VLSI device with 14–20 mask levels, requires the resolution and overlay accuracy capability of a stepper in only half of the masking levels. The remainder can be accomplished with older, more established aligner systems. Thus, less critical layers can be exposed by full-field 1X aligners, while critical layers are exposed by steppers. Increased throughput capabilities and lower cost of scanning aligners compared to steppers also reduces operating costs.

An important issue is the relative prealignment capabilities of the two systems and the time necessary to find global-alignment marks. The control of distortion is a function of the field size of the stepper.

The larger the field size, the more difficult it is to match a projection aligner to a stepper because individual distortions vary. This requires field-by-field alignment capability of the stepper in order to track the distortions of the projection aligner from die to die with tolerance thresholds at the three sigma limit. The full-field pattern must go down first, which fixes the center-to-center distances and die-to-die distortions of each row and column. Also, if the scanner produces a different horizontal magnification across the die compared to the vertical magnification, overlay accuracy is reduced.

Stepper-Stepper. A stepper mix of 5X and 10X is feasible as a mix-and-match system. In this alternative, the 10X aligner could be used for the personalization layer in master-slice type wafer fabrication due to the inexpensive production of 10X reticles. Since the 5X stepper has lower resolution capabilities, it could be used for layers requiring less registration or resolution resulting in higher average throughput. However, throughput is still lower than projection-stepper mixes, and requires higher capital investment. An additional disadvantage is the distortion that could arise by the mismatch in 5X and 10X lenses manufactured by different lens manufacturers. Maintenance problems increase when steppers from various manufacturers are mixed, since equipment is complex and scheduled and preventive maintenance routines vary among manufacturers. A mix of 1X and 5X steppers is also realistic.

Stepper-Electron Beam. A mix-and-match system that incorporates an optical stepper and an E-beam direct write system will occur as the installed base of E-beam systems increases. This approach is being used in VHSIC applications with the low-cost stepper used for non-critical masking layers and with its field-by-field capabilities used to track E-beam made levels.

The market for step-and-repeat systems will grow strongly through the mid- to late-1990s. Demand will increase as more devices are fabricated with dimensions of 0.5 µm in next-generation devices. In light of developments, optical steppers will be used for resolution of 0.35 µm for the 64 Mbit DRAM and 0.25 µm for the 256 Mbit DRAM, extending optical lithography beyond 1997.

The stepper market will be affected by advancements in E-beam and X-ray techniques, and certain devices will necessitate the need for these methodologies and impact the stepper markets within the time frame of this analysis.

5.1.5. Positive Resist Enhancement

Positive photoresists have been increasing their share of the market, largely because of higher resolution compared to negative resist. Nevertheless, increased demands for higher resolution have inhibited these advantages. Problems include:

- Differential light absorption from resist thickness variations across pattern steps
- Scattered light from neighboring patterns
- Interference between incident and reflected light

There are a number of positive resist enhancement techniques that have been developed to control the above problems as described below.

Contrast Enhancement Materials

Diffraction and focusing effects in an aligner degrade the contrast of the image, often below the contrast threshold of the resist. Contrast enhancement materials reduce the contrast threshold of the system and thus increase resolution. These materials can be applied as a thin photobleachable layer on top of the resist.

Developed into products by GE Microelectronic Materials (Phoenix, AZ), they contain dyes that bleach at wavelengths selected for the photo-optical exposure system being used.

New materials have been developed. At Mitsubishi Electric, contrast enhancement has been achieved using polysaccharides. Hitachi has developed aqueous soluble diazonium salts. Organic soluble polysilanes have also been developed into contrast enhancement products. Other companies developing products include Futurrex and MacDermid.

Anti-Reflection Coatings (ARC)

Reflections from topological features can produce notching and result in localized narrowing of the resist pattern. ARC materials, developed by Brewer Science, are organic layers containing an absorbing dye that are sandwiched between metal and resist layers. Although ARC requires a 30–50% increase in exposure energy dose, the material produces a 5–7 times increase in exposure latitude. The advantages of ARC are:

- Reduction of notching
- Elimination of standing waves
- Reduction of linewidth variations over steps.

Another method of reducing reflection is the addition of an absorbing dye to the positive photoresist. Available from American Hoechst - AZ Photoresist Products and MacDermid, they provide an intrinsic, single layer approach.

Image Reversal

Positive photoresist image reversal is a new exposure technique that renders negative toned imagery from a positive toned resist. This method has a number of advantages including:

- Greater resolution and processing latitude
- Elimination of complementary masks for both positive and negative imaging
- Higher thermal stability
- Reduction of thermal waves

Image reversal can be induced by two major methods:

- Photoresists specifically developed to reverse. Specifically designed novolac resists such as MacDermid Hi-Spec and American Hoechst AZ 5214. This method has the following disadvantages:
  — The products have a low shelf life
  — Image thickness and CD control is poor, and inconsistent image reversal sometimes occurs
- A process that exposes positive photoresist to amine vapors. The ST-A-R image enhancement process, developed by Genesis (Santa Clara, CA), requires a flood UV exposure that renders the previously unexposed regions of the photoresist layer soluble in aqueous-based developed. This method has the following disadvantages:
  — Expensive microprocessor-controlled vapor phase systems are required
  — Two additional processing steps are required

Post-Exposure Bake

Post-exposure bake (PEB) has been used for several years to reduce standing waves. Typical conditions for PEB are a soft bake before exposure followed by PEB at 90–110°C, and then development. Increasing the bake temperature to 150°C using AZ 4000 positive resist (high temperature post exposure bake or HTPEB) has been found to have the following advantages:

- Improved control of standing waves
- Improved contrast
- Thermal stability to the HTPEB temperature
- Greater processing latitude

LITHOGRAPHY

Stabilization

High vacuum or high temperature processing in plasmas, ion implanters, and deposition systems cause positive resists to degas, flow, or reticulate. Stabilization techniques have been developed to minimize these problems. Flood DUV exposure has been the accepted method, producing a hardened skin around the resist image. The process is a function of wavelength; wavelengths less than 280 nm produce a hardened skin, while wavelengths between 280–320 nm penetrates deeper into the resist.

Several new processes have been developed as alternatives:

- Electron beam curing, fostered by Electron Vision, uses a flood electron beam to cure the resist. The process is faster than DUV and does not heat the wafer.
- Photomagnetic curing from Xenon Corp.
- An aqueous solution under development by MacDermid whereby the solution, spun on top of the resist and baked, enables the resist to withstand temperatures to 175°C.

Table 5.2 presents a summary of resist enhancement methods.

5.1.6. Photoresist Materials

As exposure wavelengths become shorter for enhanced resolution, the photoresist must be optimized for sensitivity of exposure to these wavelengths. They must also withstand an increasing level of severe environments such as in plasma etching, ion implantation, and diffusion. At shorter wavelengths, conventional sensitizers for positive photoresists are no longer effective, and all components of the photoresist must be optimized for enhanced performance.

Further complicating the formulation process, submicron feature sizes are dictating additional considerations:

- Low particulates
- High purity
- Low-toxicity solvent systems
- Metal-ion-free developers

Research is currently under way at a variety of facilities—commercial resist manufacturers as well as users of lithographic tools. At IBM, for example, researchers have developed an acid catalyzed resist for an excimer laser stepper that is capable of a resolution of 0.25 µm. The deep-uv resist, XPR,

Table 5.2. **Positive photoresist enhancement methods.**

Company	Positive photoresists	Enhancement materials	Enhancement equipment	Stabilization systems
American Hoechst	o	c		
Applied Process Tech				o
Aspect Systems	o			
J. T. Baker	o			
Brewer Science	a, d			
Convac GmbH				o
Dynachem	o			
Dynamit Nobel		d		
EM Chemicals	o			
Eastman Kodak	o			
Eaton				o
Electron Vision			g	
Fusion Semiconductor			f	
Futurexx	b			
GCA				o
GE Microelectronic		b		
Genesis			e	
Headway Research			o	
Hybrid Technology			f	
KTI Chemicals	o			
MacDermid	o	b, c		
Machine Technology				o
Mallinckrodt		a, d		
Nikon Precision				o
OAI			f	
Olin Hunt	o	d		
Semiconductor Systems				o
Semix				o
Shipley	o			
Silicon Valley Group				o
Solid State Equip.				o
Solitec				o
Spectrum Resist	o			
UBC Electronics		c		
UVP		f		
Ultratech				o
Veeco Instruments				o
Xenon			h	

[a]Anti-reflection coatings, [b]Contrast enhancement materials, [c]Image reversal resists, [d]HMDS alternatives, [e]Image reversal process, [f]UV stabilization equipment, [g]E-beam stabilization equipment, [h]Photomagnetic stabilization equipment

LITHOGRAPHY

uses a chemical amplification scheme for improved sensitivity. Exposure decomposes a photoactive precursor to yield an acid moiety which, in turn, modifies the solubility of a proprietary resin during exposure. A small amount of a photoacid is produced, increasing the alkaline solubility of the resin in exposed areas, and affecting a large change in resin solubility and sensitivity.

Listed in Table 5.3 are characteristics of currently available commercial photoresists.

5.2. ELECTRON BEAM SYSTEMS

As described above, the shorter wavelengths of the illumination source has extended the resolution of optical systems to the sub-0.5-µm-linewidth region. The 4 to 5 times decrease in the wavelength of electrons compared to optical (visible and ultraviolet) radiation permits resolution as small as 0.2 µm. In addition, the charged electrons can be easily rastered and vectored by electrostatic and magnetic fields as illustrated in Figure 5.3 (Malmberg et al. 1973) (Thomson et al. 1987). This facilitates faster movement of the beam and more flexibility in writing a variety of patterns on a mask, reticle, or wafer. Optical photons, on the otherhand, are mechanically moved by mirrors or by movement of the stage.

The following sections describe the technological issues and trends in the use of E-beam lithography for mask making and direct write applications.

5.2.1. Mask Making

The most important issue impacting mask making E-beam systems is registration accuracy. Since there are no registration marks on masks, as opposed to wafers, the referencing must be within the system itself. This places stringent requirements on the mechanical stability of the E-beam system. A single layer written across a 6-inch mask requires a high degree of stability to maintain registration accuracy as no interim realignment is performed. In order to achieve a 0.1 µm overlay accuracy, precise temperature control and table positioning are required. A one degree change in temperature can cause misregistration in two successive glass mask levels as large as 0.25 µm over a 3-inch area. Use of fused quartz could reduce misalignment to 0.1 µm over a 4-inch area.

Sophisticated E-beam systems, selling for as much as $4 million, utilize the computer-aided-design (CAD) pattern output to generate the masks and reticles. Software capabilities of the E-beam computer are used to control writing functions such as stage motion, beam deflection, and blanking speed.

A problem with E-beam-generated masks and reticles is roughness of mask edges, particularly for 1X masks. This has been reduced in recent years

Table 5.3. Commercial photoresist characteristics.

Company name/Model	Type	G	H	I	Deep UV	E-beam	X-ray	Resolution limit (μm)
J. T. Baker								
1-PR Series	+	o						0.5
E38	+			o	o			0.3
Dynachem								
EL 2015	+	o	o	o				0.5
EL 2025	+	o	o	o				0.5
EPA 914 EZ	+	o	o					0.6
NOVA 2050 AR–2	+	o	o					0.6
NOVA 2070	+	o	o	o				0.35
OMR–83	–	o						
Fuji-Hunt								
FH-6450	+	o			o			0.55
FH-EX1	+							0.35
Hoechst Celanese								
AZ 1300 Series	+	o						
AZ P4000 Series	+	o						
AZ 5200 Series	+		o	o				
KTI Chemicals								
732	–				o	o	o	1.5
820	+	o	o	o	o	o	o	0.5

LITHOGRAPHY

895i	+			o		0.5
PMMA	–					0.01
MacDermid						
Ultramac EPA 914	+	o				0.5
Ultramac PR 914	+	o				0.5
Ultramac PR 1024	+		o		o	
OCG Microelectronic Materials						
HEBR-214	+		o	o	o	0.2
HiPR 6500	+		o	o		0.45
Hr-100,200,300	–		o	o		2.5
Selectilux P3100	+		o	o		0.5
Shipley						
Megaposit SPR500	+		o	o		0.4
Megaposit S1800	+	o				0.8
Megaposit SNR248	+			o		0.25
Sumitomo Chemical						
Sumiresist PFI–15	+		o			0.45
Tokyo Ohka						
THMR–iP1800	+		o			0.35
Toray						
PR–a12—	+	o				0.38

96   **LITHOGRAPHY**

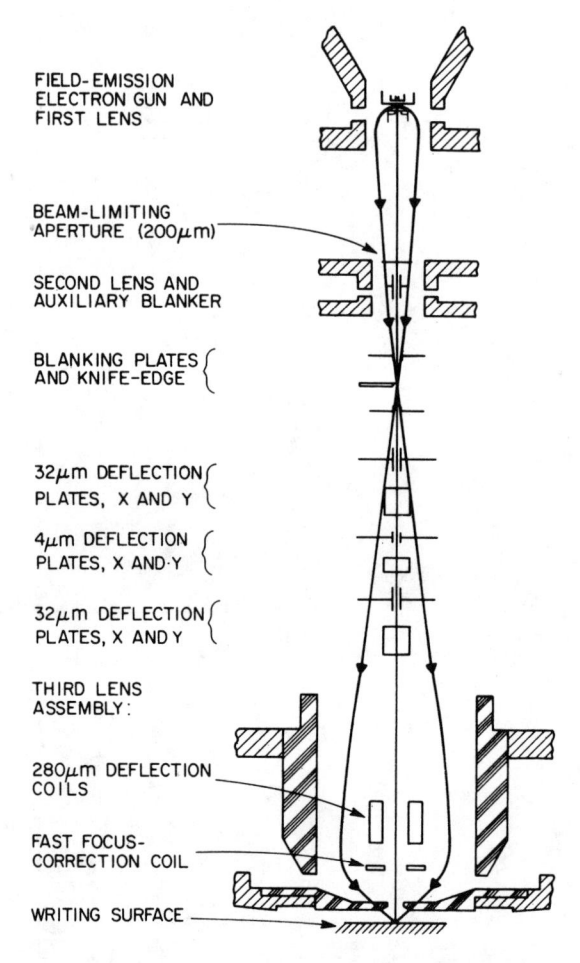

Figure 5.3. Schematic of electron beam system.

by improved electronics. A significant improvement has been attained by generating a 10X reticle with an E-beam and then using a conventional step-and-repeat camera to produce the 1X mask.

The major technological advantage of E-beam mask-making systems compared to optical techniques is its increased speed and better accuracy. Complex circuits that would require over 10 hours to generate on an optical pattern generator can be made in less than an hour with an E-beam system. In addition, overlay accuracy of 0.1 µm can be attained with an E-beam compared to 0.4 µm with an optical pattern generator. However, edge roughness can still be a problem if a raster scan rather than a vector scan system is used.

5.2.2. Direct Write

The high quality reticles generated by E-beam systems have contributed to the success of the optical projection and stepper aligners and indirectly slowed the use of direct-write E-beam systems. The overriding issue in direct write E-beam has been throughput. Until recently, throughputs as low as only 10 wafers per hour could be attained. New-generation systems can now offer a throughput of 30, 4-inch wafers per hour on a 1 µm minimum linewidth wafer for the last three personalized metal layers.

Throughput of a system depends directly on:

- Data preparation
- Data verification
- Data manipulation
- Tool throughput
- Yield
- Inspection and repair time

As opposed to mask-making systems, direct write registration is achieved by referencing to registration marks on the wafer. The beam is re-registered after each die is written so that long-term system stability is not critical.

The primary applications for direct-write, E-beam systems are for:

- High-density circuits that require high resolution and overlay accuracy, such as GaAs or VHSIC devices
- Quick turnaround production, such as ASICs.
- Research and Development

Submicron feature size is the norm on microwave (MMIC) GaAs devices. At the present time, these devices are primarily at the MSI level of component density—some as few as 10 components. The few features are generally not affected by particulates coming from an optical contact mask. Direct-write E-beam lithography will eventually be implemented by most of the GaAs IC manufacturers. It is already installed by most of the large GaAs manufacturers, but the GaAs industry is largely composed of small startups. As integration goes to the VLSI levels, more purchases will be made.

ASIC devices need not only an accurate lithographic system, but a quick one since, unlike mass produced memories, they tend to be required in comparatively small quantities over a shorter time span. The typical number of ASIC devices required for a given application is slightly less than 5,000 and decreasing. In these applications, the speed of the E-beam system is as important as its accuracy. For ASICs, minimum feature sizes of 0.8 µm are well within the limits of optical steppers.

Nevertheless, the prohibitively high cost of masks—$1,000 to $1,500 prior to inspection and repair—will make direct-write E-beam technology cost effective. Also, the slowness of the system is counter productive (Wilson 1986). The slowness of E-beam machines is inherent in the E-beam process, but attributed to the decreasing size of the grid structures that designers use to lay out the mask pattern as well. The smaller the grid, the longer the processing. Thus, although 0.1 µm grids are essential, larger grids should be used for less critical portions of the design.

Another problem responsible for the slow speed of direct write E-beam is resist quality in production environments. (This resist issue is also a problem with mask making E-beam lithography, but throughput levels are not critical).

Several direct write E-beam production companies that projected high sales volume a few years ago dropped out of the market. Companies such as Control Data, GCA, General Signal, and Veeco recognized the small market potential. Of the two remaining U.S. suppliers, Varian exited the lithography marketplace because of problems in developing its direct-write system, the VLS-1000. Etec (previously Perkin-Elmer's Electron Beam Technology division and owned by IBM, Grumman, DuPont, Micron Technology, and Zitel) also experienced delays, and it's AEBLE-150 was late into the marketplace. JEOL's JBX-5DII system has been warmly welcomed in the U.S., including the National Security Agency, the Naval Research Laboratory, Massachussetts Institute of Technology, Cornell University, University of Michigan, Jet Propulsion Laboratory, and Wright-Patterson Air Force base.

Several JBX-6A systems have been installed in the U.S. (several are installed in Honeywell) where it is used for both mask making and direct write. Etec's success in getting the AEBLE-150 to the market reduced sales of JEOL's system.

Significant improvements in some E-beam systems have moved these systems into consideration for production of ICs:

- Concentrating on high throughput, Etec's variable-shaped-beam AEBLE 150, designed for the VHSIC program, has a throughput of 30 wafer levels/hr for ICs with 0.5-µm feature sizes to an overlay accuracy of 0.1 µm, and a critical dimensional control of 0.05 µm. P-E's software problems and difficulty in obtaining the proper acuity of lines and registration of patterns on the wafer have been resolved, and several have been delivered for VHSIC chip production, Raytheon for GaAs manufacture, and to European Silicon Structures, France, for ASICs.

- Cambridge Instruments (now Leica), with an installed base of nearly 60 systems, supplies a direct-write system for sub-0.1-µm linewidth for GaAs FETs and MMICs, and a system for ASIC devices. Recently, the company has received orders for a Gaussian beam EBMF 10.5 from Varian and Gain Electronics for use in GaAs devices.

- JEOL's JBX-6A11, with a variable shaped beam, has a speed of 10 wafers/hr and an accuracy of 0.1 µm. Thermal expansion of the mask plate during processing is a problem that has been overcome by the use of three independent temperature controllers. The JBX-5II, with a Gaussian beam, is used extensively for R&D and ASICs.

- ASM Lithography's Beamwriter EBPG-4 Gaussian beam system has been designed for direct writing onto silicon and GaAs wafers, allowing linewidths down to 0.1 µm. Recent orders have been placed by Avantek, Hughes Research, and Microwave Semiconductor Corp.

- Hitachi's HL700D uses a variable beam at a usable beam current density of only 5 A/cm^2.

- Lepton's $6 million EBES4 family can produce up to five reticles per hour with feature sizes as small as 0.25 µm. The system can also write directly on silicon wafers with a throughput for up to five 8-inch wafers per hour with feature sizes of 0.4 µm and smaller.

Advances in optical techniques and anticipated advantages in X-ray lithography have been a mixed blessing for E-beam systems. E-beam mask making systems are proving to be the best method for mask and reticle generation for the large installed base of optical aligners and steppers and for X-ray masks. On the other hand, optical methods are capable of 0.35-µm features, which are not anticipated to be used for production devices for several years. By that time, the technological problems associated with X-ray steppers should be resolved.

The technological problems such as slow speed and proximity effects that are associated with direct-write systems will not be a factor for certain devices such as ASICs, and VHSICs, as well as some GaAs. This issue will also be compensated by the increased use of mix-and-match approaches combining steppers and E-beams.

An important development in captive E-beam systems was the disclosure in 1986 that IBM's EL3 and ELX (renamed the EL4) systems would be manufactured by Grumman, a variable shaped beam system. Up to 50 systems were to be built for IBM's use in bipolar logic and MOS VHSIC devices.

IBM announced in April, 1990 that Etec will build the newer version, the EL4. The organization includes Grumman, with expertise in building the EL3 and EL4, and DuPont, the largest supplier of photomasks.

The contract will have a number of far-reaching implications:

- IBM, a large purchaser of process equipment, will minimize the need to purchase from the outside any E-beam systems to meet internal requirements. DuPont will be a big user in their worldwide mask shops.

- Etec will commercially market the system as a replacement for its aging MEBES III line, in which only two systems were sold in 1989.

- The EL4 will become a successor to the Able-150, under modification to write submicron features.

There are several additional areas that need improvement for E-beam to continue to generate sales volume (Berglund 1989):

- Resist sensitivity is at an early stage. Optical resists exhibit sensitivities on the order of 20 to 40 uC/cm^2. This seriously limits the throughput of E- beam. Although advances in negative E-beam resist have dropped sensitivities to 2 to 5 uC/cm^2, positive resists are not yet available.

- Proximity effects in single-level resist become increasingly problematical below 1 μm. With multilevel resist, proximity becomes a problem at 0.5 μm. Multilevel resist processing adds to the complexity of device processing. Proximity correction of the data in software also requires a significant data preparation overhead.

- The current number of flashes at each wafer level is too high. The number of flashes depends on the design methodology and data preparation. Improvements on computing hardware and software tools and optimized design methodology will reduce the flash count.

- A dwell time of 100 ns of each flash is too high. This can be minimized once resist technology has improved and by further improvements in machine design.

Based on IBM's research on projection E-beam in the early 1980s, AT&T Bell Laboratories is working on feasibility studies for projection E-beam lithography. Called Scalpel (SCattering with Angular Limitation Projection E-beam Lithography), the company has already printed 0.1-μm features in a 1 mm^2 field. The system uses a very thin non-absorbing mask, with circuit patterns that scatter electrons from the beam line. After passing through the lens, the scattered electrons are filtered out, leaving only those not affected by the mask to reach the wafer.

LITHOGRAPHY

Table 5.4. Characteristics of X-ray systems.

	Conventional	Plasma	Synchrotron
Generation mode	Electron bombardment of water-cooled source	Laser or spark induced plasma discharge	Storage ring
Flux (mW/cm^2)	0.005–0.30	1–15	100–1000
Feature size	0.5 µm	0.3 µm	0.1 µm
Wavelength (nm)	0.4–1.0	1.4	0.6–1.0
Field size (mm)	40×40	20×20	50×50
Problems	Low flux, Short wavelength	Unreliable, Contaminates windows	Very expensive

5.3. X-RAY SYSTEMS

5.3.1. X-Ray Sources

X-ray lithography utilizes a proximity printing process similar to methods described in Section 5.1.1. Soft X-rays with a characteristic wavelength of 0.4 to 1.0 nm are employed. Several types of X-ray sources are used, and the various methods are described in Table 5.4 and illustrated in Figure 5.4 (Branst, 1987).

For conventional electron bombardment and gaseous or metallic plasma sources, a typical system incorporates an X-ray-generation high vacuum chamber with a 25-50 µm thick beryllium window, a helium column separating the X-ray source and mask, and a proximity chamber separating the mask and substrate. In the plasma source, a high-voltage pulse is sent through a gas sample that forms an intense plasma that radiates a burst of soft X-rays. This source has an X-ray efficiency up to 100 times higher than the 0.01% of a conventional X-ray tube, as well as a spot size of 1.5 mm compared to 4 mm (Heuberger 1986).

There are three types of lasers (Nagel *et al.* 1978) (Branst 1987) used to generate X-rays: excimer, Nd:glass, and Nd:glass slab. Focusing a laser onto a metal target produces a plasma that is composed of ions, free electrons, and neutral atoms. The neutral atoms interact with the high-peak-power laser pulse and result in multiphoton absorption and highly excited bound electrons. The free electrons combine with the ions, and excited electrons decay

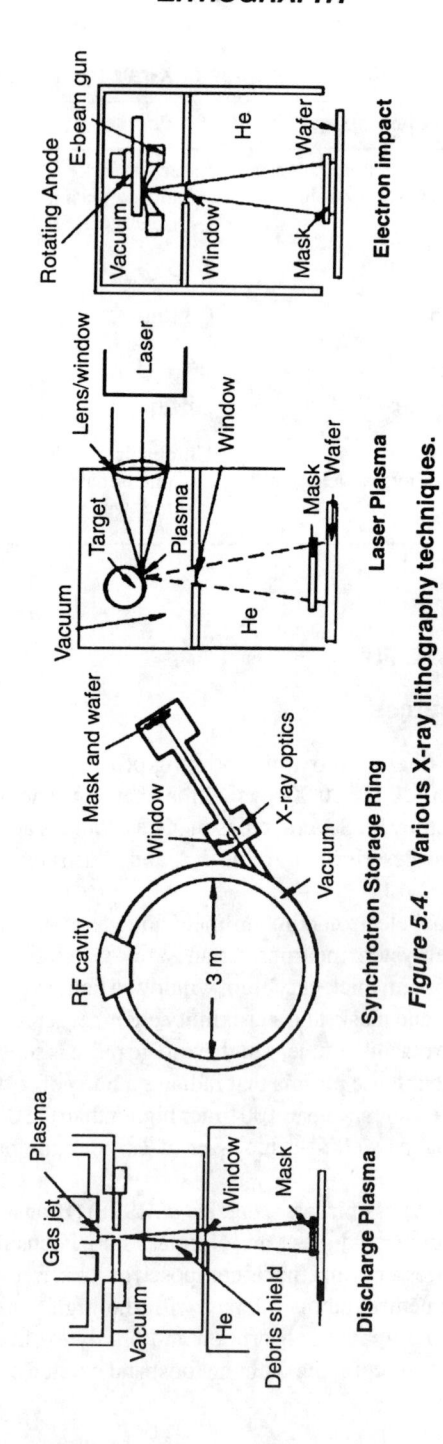

Figure 5.4. Various X-ray lithography techniques.

LITHOGRAPHY 103

to lower energy levels to produce X-rays. As much as 20% of the laser energy can be converted into X-rays in an efficient plasma source compared to 0.01% in a conventional X-ray tube. The spot size is also smaller; 1.5 mm compared to 4 mm.

- An excimer laser, used in UV and DUV sources described above, present a safety hazard in the fab area. Excimer lasers must be filled with a toxic gas (fluorine or hydrogen chloride) continuously during the day through separate cylinders of gas. Also the heads must be tuned frequently and rebuilt once a month.
- A Nd:glass laser, operating in the infrared at 1060 nm, is safer and less expensive to maintain than an excimer laser. There is also no need to shield the wafer from scattered laser radiation; UV-generated sources require windows or filters to minimize DUV laser light reaching the wafer, reducing X-ray intensity at the wafer.
- A Nd:glass slab laser uses a glass slab as an amplifier in which the laser beam is amplified while passing in a zig-zag course through the slab. The slab is simpler and requires fewer optical elements than rod amplifiers. However, it is prone to thermal gradients that can cause cracking.

Critical parameters to achieving high resolution are:

- The mask-to-substrate gap
- Target-to-substrate distance
- Emission source diameter

These factors control the penumbral blur that broadens the transmissive areas of the mask. The collimated resolution from a synchrotron (Spiller and Feder 1977) source prevents penumbral blur. However, synchrotron sources are proving to be too expensive ($16 million) for extensive commercial use at this time. Most applications are directed toward research use in which a facility can be located in close proximity to a synchrotron radiation source in order to tap off the X-rays. This method is illustrated in Figure 5.5 (Heuberger 1986).

Other factors affecting resolution include:

- Secondary electron range
- Mask transmission and line-edge profile
- Resist properties
- Target or anode source

The latter is of importance, since the emission spectrum of the source must be matched to the adsorption characteristics of the transfer media, mask,

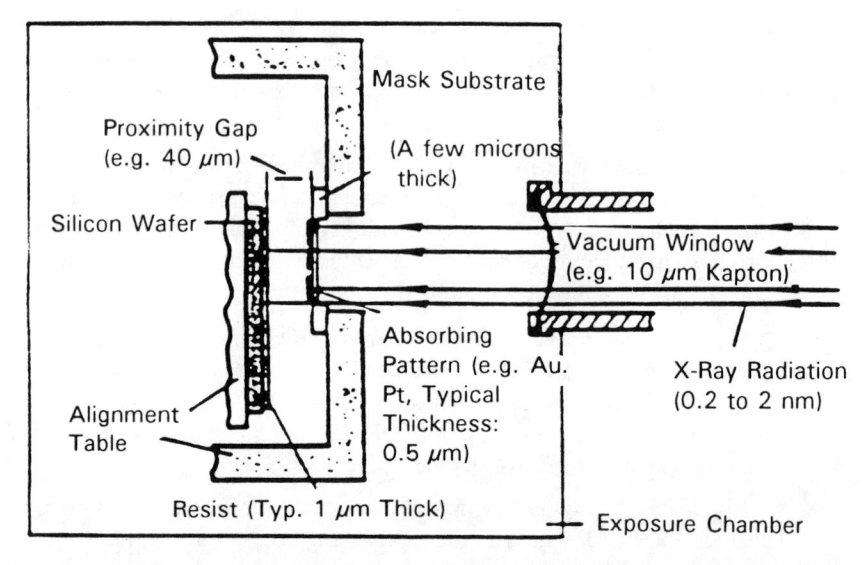

Figure 5.5. Schematic of X-ray lithography using synchrotron radiation.

and resist in order to achieve high exposure rates and exposure contrast. For electron bombardment and plasma sources, the most commonly used X-ray targets include aluminum (Al), copper (Cu), molybdenum (Mo), palladium (Pd), rhodium (Rh), silicon (Si), and tungsten (W) (Lingnau et al. 1989). Source-mask combinations that have shown the highest performance are palladium-boron nitride, and silicon or aluminum with silicon carbide or silicon nitride masks. An advantage of palladium is its efficiency in passing through organic dust particles and ease of blocking by a thin gold absorber.

The X-ray source requires input powers greater than 500 W, necessitating the use of stationary liquid-cooled targets. Targets are typically cone shaped (25°) and cooled by high-pressure (200 lbs/in.^{2}), high-purity water.

The U.S. has established a limited National X-ray Lithography Program to establish a 0.25-µm IC manufacturing process. The Program has received $35 million ($15 million in 1988 and $20 million in 1989) in federal funding. Although a supplier of the actual X-ray system has not been decided, the Program has already awarded Micrion a $1.2 million contract for a specially configured version of the company's Model 808 focused ion beam mask repair system. Additional mask-related contracts have been awarded to Stanford University, MIT, Rohm & Haas, and Spire. New membrane and absorber material development is being coordinated with Sematech.

LITHOGRAPHY

The House Armed Services Committee doubled funding for X-ray lithography to $40 million for Fiscal 1990. This contrasts with more than $1 billion that IBM alone has spent toward development of an X-ray facility.

There are several active areas of research into X-ray systems in the U.S.:

- Perkin–Elmer (now SVG Lithography) has developed an X-ray stepper, the XSAR2, for its VHSIC participation that currently uses a conventional rotating anode source that could eventually be replaced by a plasma source. A system has been shipped to the Solid State Electronics Division of Honeywell. The subsystem design characteristics include:
  - Source: Electron impact, 8000 rpm, 10 kW tungsten anode, 1.5 mm spot
  - Mask: 4-μm boron nitride, 0.7-μm Au absorber, >50% X-ray transmission
  - Alignment: Linear zone plate-grating HeNe sensing, capacitor gap sensing
  - Resolution: 0.5-μm lines/spaces, ±0.05 μm
  - Overlay: ±0.1 μm
  - Throughput: 20, 100-mm-diameter wafer levels/hour
  - Reliability: 90% uptime for 24-hour operation

- Hampshire Instruments introduced the Series 3500 X-ray stepper, a laser-based system capable of sub 0.35-μm features. The stepper, which replaces the company's earlier Series 5000 models, utilizes the spectrum of "soft" X-rays, allowing the use of conventional novolac resists. The system uses proximity printing, in which a 1:1 reticle is held close to the wafer and point-source X-rays are passed through it. (Peters *et al.* 1988) (Preston *et al.* 1989).

- AT&T, one of the pioneers in X-ray lithography, has developed a water-cooled electron-impact palladium source, boron nitride mask technology, and high-sensitivity X-ray resists. Currently on its third generation full-field system, the company is directing efforts on understanding pattern distortions caused by stressed X-ray absorber features as well as the stability of mask materials that are exposed to X-rays from a storage ring. The projection X-ray system could be capable of 0.1-μm features. AT&T has purchased several 5000P X-ray steppers from Hampshire.

- IBM has been a participant in the National Synchrotron Light Source vacuum ultraviolet ring project at Brookhaven National Laboratory. IBM has been the principle user of one of the ports on the VUV storage ring and has designed a vertical stepper (as opposed to the Micronix horizontal stepper in which wafers are mounted horizontally) to produce several prototype devices. IBM is undertaking extensive work on membranes and absorbers to produce distortion-free masks using silicon and boron nitride additively patterned types.

IBM also has installed an X-ray stepper purchased from Karl Suss America to an IBM installation at Brookhaven.

IBM, in early 1987, signed a $16 million contract with Oxford Instruments (UK) to build a compact ring, that was installed in 1991 at IBM's East Fishkill, N.Y. facility.

IBM received a $17.4 million contract from the Naval Research Laboratory's National X-Ray Lithography Program to work on X-ray masks.

IBM dedicated its Advanced Semiconductor Technology Center in East Fishkill, NY in late 1989. It will house the first privately owned synchrotron storage ring for X-ray lithography in the U.S. Motorola was the first company to announce collaboration with IBM to develop technology to achieve sub-0.25 µm features.

- Lawrence Livermore National Laboratories is developing an X-ray technology using a conventional source.

- The Department of Energy has funded the construction of a synchrotron at Louisiana State University.

- The University of Wisconsin Synchrotron Radiation Center has established the Wisconsin Center for X-ray Lithography. Two beam lines in the storage ring at the university, called Aladdin, are dedicated to research in X-ray lithography, including mask fabrication, mask distortion and damage assessment, oxide radiation damage assessment, resist characteristics, optics, beam line design, and process modeling. SVG Lithography will modify its XSAR2 stepper for use on Aladdin.

- Naval Research Center, with DARPA (Defense Advanced Research Projects Agency) backing, is negotiating to set up a dedicated X-ray lithography line.

- Grumman is working under a DARPA contract awarded to Brookhaven National Laboratory to develop a compact superconducting synchrotron, which could enter the commercial marketplace by 1994. Grumman will then market the commercial system, the Superconducting X-ray Light Source (SXLS), a 10-angstrom source similar to Oxford's Helios.

In Japan, efforts on X-ray lithography are directed on several main fronts:

- Conventional lithography systems are being manufactured by Nippon Kogaku (Nikon) and Canon that are commercially viable for semiconductor production. Mitsubishi fabricated a 1 Mbit DRAM using a Nikon X-ray stepper that employs a KrF excimer laser.

- Sumitomo Heavy Industries and NTT have begun manufacturing prototype industrial-use synchrotron orbital radiation (SOR) rings. A prototype of the Aurora compact ring (1-meter diameter) became available in late 1989. The wavelength is 1.02 nm at an energy of 650 MeV, with an irradiative power of 1.5 W/mrad at a stored electron current of 300 mA. Chip fabrication began late 1990. The machine sells for between $13 and $17 million.

- Fujitsu and NEC have developed prototype SOR steppers. Fujitsu has developed the technology for the production of 0.3-µm feature sizes for use on 64 Mbit DRAMs. NEC is using an open-air SOR and has obtained feature sizes of 0.2 µm.

- NTT's prototype is an 8 foot by 28 foot machine that weighs more than 60 tons. Called Super-ALIS, it is being used for chip fabrication at NTT's LSI laboratory in Atsugi City, Kanagwa.

- At the Electrotechnical Laboratory operated by the Ministry of International Trade and Industry, Sumitomo Electric Industries completed Niji-III in early 1990, culminating a 5-year effort. The system is priced at $14 million.

- A group of 13 Japanese electronics firms has formed Sortec, a 10 year, $100 million consortium to develop an X-ray lithography system based on an advanced synchrotron source. Eight compact rings are in various stages of planning, and one is nearing completion. Sortec membership includes: Canon, Fujitsu, Hitachi, Matsushita, Mitsubishi, NEC, Nikon, Oki, Sanyo, Sharp, Sony, Sumitomo Electric, and Toshiba. Seventy percent of the funding was made by Japan Key Technology Center under the jurisdiction of MITI.

- NTT has organized a program for the development of all aspects of X-ray lithography, including both conventional and synchrotron sources. Over $50 million has been committed to the program. Additional research is being conducted by NEC, Fujitsu, Hitachi, Nikon Kogaku, the national laboratories, and universities.

In Europe, efforts underway are:

- X-ray lithography technology started as a multicompany, government-supported project at the Fraunhofer Institute for Microstructure Technology, West Berlin. The near-term goal was to use synchrotron radiation to produce 0.5-µm structures as well as a commercial X-ray stepper in 1987. Researchers at the Fraunhofer Institute have developed a practical synchrotron called COSY, which has attained 0.2-μm pattern sizes in an R&D mode. Using the system, engineers from Telefunken Electronics GmbH have fabricated a dual-gate NMOS FET for operation up to 1.2 GHz. The chips have 1-μm gate lengths and measure 0.5 by 0.5 mm. MOS FETs with 0.5-μm gate lengths were fabricated in 1987. The technology had been licensed to Leybold-Heraeus, whose startup COSY MicroTec was to have marketed a compact synchrotron by 1990. However, because COSY is three years behind schedule, COSY MicroTec was disbanded and Fraunhofer is looking for another marketeer.

- Siemens, Philips, and SGS-Thompson are considering continuing the Cosy Microtec program as part of their Joint European Submicron Silicon (Jessi) research consortium that is still in the planning stages.

- Under the ESPRIT program, a consortium of Thompson, Imperial College, and SGS-Thompson are installing a step-and-repeat alignment system on the storage ring at Frascatti, Italy.

- Karl Suss has developed the XRS-200 stepper that uses either a synchrotron or plasma-focus source. The system, priced at $1 million without source, is capable of 0.2-μm resolution and 0.1-μm alignment accuracy over fields ranging from 25 × 25 mm and 45 × 45 mm. Throughput is specified at 20 6-inch wafers per hour. Companies using the earlier MAX 1 stepper include Siemens–Sietec, Philips–Valvo, AEG Telefunken, IMT, IHT, and Eurosil Eching.

- A superconducting synchrotron ring for X-ray lithography will be constructed by Scandtronix of Uppsala in Sweden.

In the People's Republic of China, efforts towards X-ray lithography research are on two fronts:

- The Beijing Electron Positron Collider (BEPC) located at the Institute of High Energy Physics has five beams dedicated to research work including lithography.

- The Hefei National Synchrotron Radiation Laboratory is located at the University of Science and Technology of China, Hefei, Anhui Province.

Much of the national work on X-ray lithography focuses on synchrotron sources. Characteristics of these sources shown above in Table 5.4, include:

- High power
- High collimation

LITHOGRAPHY

- Small effective source size (high brightness)
- Broad band, tunable, incoherent

These characteristics give rise to the following advantages of using synchrotron X-ray lithography:

- High throughput
- Low cost per wafer exposure
- Process simplicity
- High aspect ratios
- High resolution
- Compatibility with X-ray assisted in-process monitoring and low-temperature processing

Listed in Table 5.5 are the current storage rings and planned construction in the world. X-ray lithography rings are optimized for operation in the 0.5 to 1.5 GeV range. The SPEAR and PEP rings at Stanford and the X-ray ring at Brookhaven operate at energies too high for X-ray lithography.

5.3.2. Mask Making

Mask making has been the key reason for the late introduction of X-ray lithography systems in fab lines. A typical mask is composed of a thick layer of a high absorbing material (e.g., 0.7 μm of gold) on top of a thin, low density, transparent membrane or pellicle (Mauger and Shimkunas 1986). Substrate materials under investigation include boron nitride, silicon nitride, silicon carbide, silicon, polyimide, and foils (Hampshire uses silicon and silicon carbide membranes with tungsten absorbers). Polyimide is used to absorb photoelectrons ejected from the gold pattern, provide shock resistance to sudden impact, and tolerate local distortion. Gold and molybdenum are the common absorbers (Smith *et al.* 1973) (Maydan *et al.* 1975).

Boron nitride mask technology is adequate for 0.5-μm features. The lifetime of the mask can be improved by changing the membrane materials. Membranes made using LPCVD with diborane and ammonia become damaged at synchrotron-produced radiation of 200 kJ/cm^3. Membranes using hydrogen-poor borazine are not damaged.

Silicon nitride has similar properties to boron nitride and is used with gold and tungsten absorbers. Mass-produced masks from NEC have been made with consistent properties. Silicon is 100 times more resistant to synchrotron radiation than the other two materials. It is feasible that this material

Table 5.5. Worldwide synchrotron-based XLR.

Country	Existing rings	Existing XLR programs	Rings in development	Planned new XLR programs	Dedicated XLR XLR (total)
Brazil			1	1	
China	2	2	1	2	
England	1				
France	2	2	1	2	
Germany	3	2	2		2
India			1	1	
Italy	1	1	1	1	
Japan	6	3	7	10	8
Korea			1	1	
Sweden	1		1	1	
Taiwan			1		
USA	7	2	4	4	2
USSR	5	1	1		2
TOTAL	28	13	22	24	14

can be fabricated into a monolithic mask, with Si used as the membrane and absorber. As a result, this material will replace boron nitride and silicon nitride.

Silicon carbide has the greatest strength of all the materials, and with gold as an absorber, exhibits better distortion and stability characteristics than boron nitride and silicon nitride. This material is currently used by Fujitsu to fabricate 0.25-µm geometries.

In Japan, NTT has demonstrated mask fabrication at 0.2 µm, while Dai Nippon Printing now makes masks at 0.3-µm.

Electron-beam pattern generators are used to define the submicron masks. However, proximity affects are significant if the masks are written with the heavy metal absorber already in place. The price of an X-ray mask is a function of the field size and address structure. A 100-mm, full-field mask with a 0.1 µm address structure costs $10,000 to $12,000 prior to inspection, and repair costs of $1,000 to $1,500. As discussed above, the price of a stepper mask with a 30×30 mm field is $3,000 to $4,000. However, X-ray masks, unlike optical masks, can be used as masters for the generation of working masks, reducing the cost of the mask to the inspection and repair costs.

The technique of X-ray lithography utilizes 1:1 masks as with optical steppers and E-beam systems. The difficulty in maintaining rigidity of such a thin membrane over a large area is a major reason for the move toward the development of X-ray steppers rather than projectors that have been employed for several years at AT&T Bell Laboratories. Also, projection aligners are adequate for only 4-inch wafers at resolutions of 1 µm. The industry trend toward large wafers is also a driving force for the use of X-ray steppers.

5.3.3. X-ray Steppers

An X-ray step and-repeat system, illustrated in Figure 5.6 (Huber *et al.* 1990) has several advantages over an X-ray projection system:

- The practical limit for a full-field X-ray system is 100 mm. Above this, mask fabrication is extremely difficult because of the fragile membrane. The prevalent use of wafers today is 125 and 150 mm diameters.

- In an analogy to optical systems, X-ray step-and-repeat systems are more accurate than full field systems. Also, a mask with a smaller patterned field, whose image can be stepped onto a wafer, has a lower distortion than a full-field mask.

In Japan, Nippon Kogaku has announced a stepper for 150 mm wafers that uses a 7.1 Angstrom wavelength source that can provide linewidths of

112   *LITHOGRAPHY*

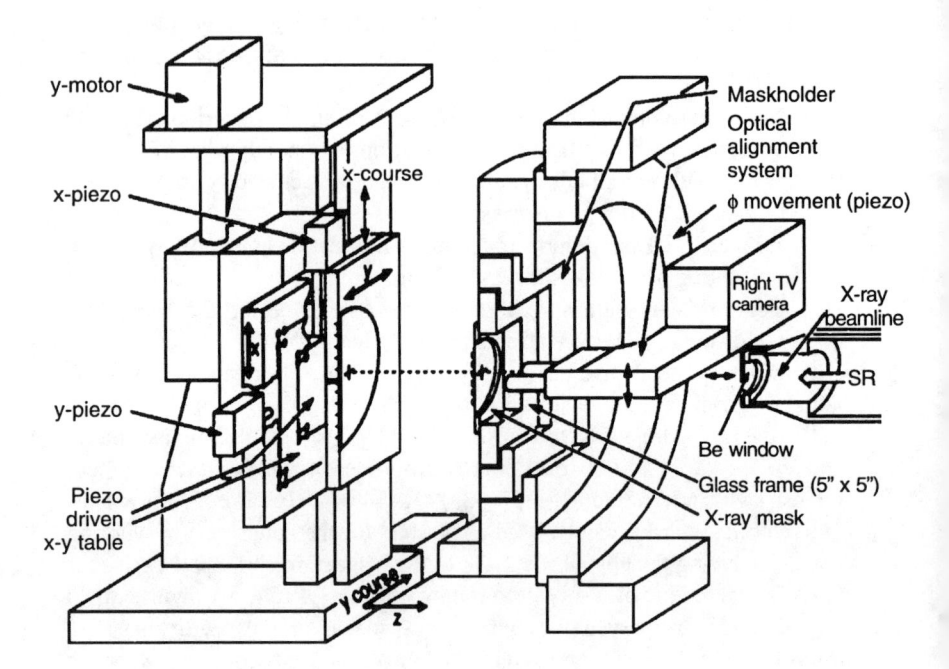

Figure 5.6. Schematic of X-ray stepper.

0.5 µm with a throughput of 4, 150 mm wafers. The exposure area is 25 by 25 mm, compared to the Micronix stepper that had an exposure field of 50 by 50 mm. The company has announced that it has sold four systems, including two to NTT and one to NEC. Canon has also introduced an X-ray stepper using 7.1 Angstrom radiation and directed for use in 16 Kbit and 64 Kbit DRAM research. Available in 1991, the system has an alignment accuracy of 0.01 µm, sufficient to fabricate 0.1-µm devices.

Shipments of X-ray systems began with the announcement by Micronix that the first MX-15 system was delivered to a semiconductor manufacturer in late 1984. Both companies worked together to develop the system. This follows the introduction by Micronix of the first commercially available boron nitride masks in May, 1984. The first commercial X-ray stepper was introduced by Micronix in 1986.

Several joint ventures and licensing agreements have been responsible for advances in X-ray lithography such as:

- AT&T Bell Labs—Micronix
- AT&T Bell Labs—Spire

- AT&T Bell Labs—SVG Lithography
- Hewlett-Packard—Varian

As many as seven merchant suppliers of X-ray systems have entered the market place, while companies such as AT&T Bell Labs, AT&T Technologies, Hewlett-Packard, IBM, Intel, National Semiconductor, and Texas Instruments are developing their own in-house capabilities.

The market will grow in the 1990s with the elimination of mask making problems brought about by the entry of new companies into this market niche. However, because of the wafer size limitation of full-field X-ray systems, market acceptance will be slow until X-ray steppers are proven production worthy. An underlying detriment to growth will be the extension of optical lithography to the 256 Mbit DRAM in the late 1990s.

Current X-ray steppers use a 1:1 proximity method. A reduction system for X-rays, in which the mask's image is reduced onto the wafer, offers the advantage of a simpler mask generation process. Also, a mask system can be thicker and stronger than the thin membranes now used.

To meet these advantages, researchers at AT&T Bell Labs and GCA's Tropel division have built an experimental X-ray projection system with a 20:1 reduction ratio. With this system, a 0.5-μm feature requires a mask with 10-μm features, rather than 0.5-μm features currently.

At the present time, features as small as 0.05 μm have been patterned on silicon. Photomasks were produced on a 0.7-μm thick silicon substrate with a gold contrast enhancement layer or a germanium absorption layer applied. A three-layer resist was comprised of PMMA, germanium, and a G-line optical resist (Shipley) to achieve the resolution.

A limitation of the system is the small field size—25 by 50 μm, which needs to be enlarged for production.

The X-ray source was Brookhaven National Laboratory's synchrotron source, which is the same source used by IBM for its X-ray program. The beam was directed through a 20X reduction Schwarzchild-type objective, using a pair of almost-concentric spherical mirrors coated with extremely thin layers of either iridium and chromium or molybdenum and silicon, depending on the wavelength. Advanced systems will use a different multi-mirror scheme. A schematic of the system is shown in Figure 5.7 (Dunn, 1990).

5.3.4. X-ray Resists

Synchrotron and laser X-ray sources emit energy in a narrow band of radiation in the 0.8 to 1.6 nm region referred to as "soft" X-rays. However, synchrotron X-rays have a "hard" wavelength component that peaks in the 0.6 to 1.0 nm range (Plotnik 1987).

114 **LITHOGRAPHY**

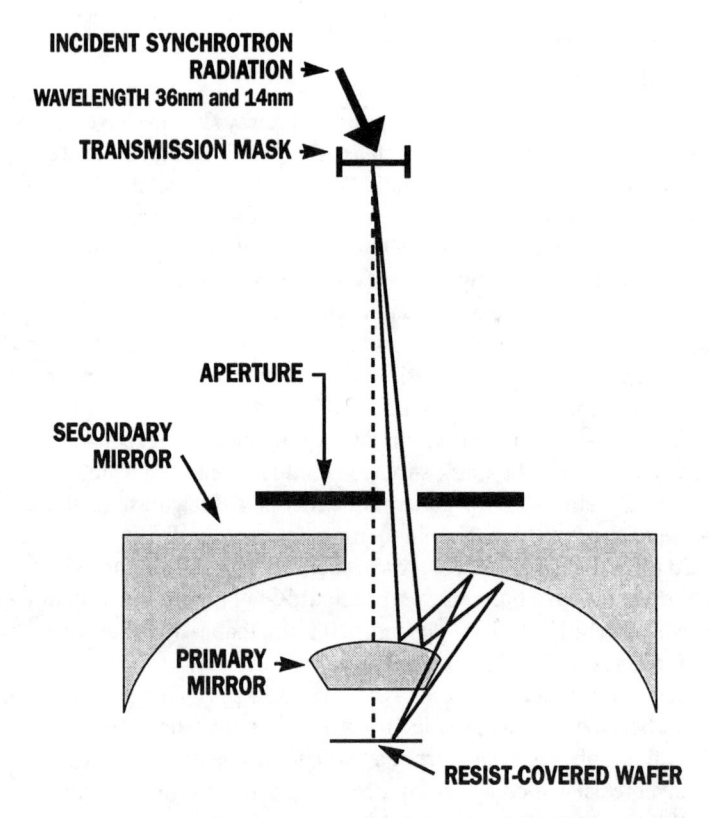

Figure 5.7. Schematic of X-ray reduction system.

Conventional novolac resists used in optical lithography are also sensitive to these soft wavelengths. At a wavelength of >1.2 nm, 40 to 60% of the radiation is absorbed in a 1-µm film of novolac resist—similar to optical mid-UV exposure. Further enhancements in speed are probable through different developer formulations and processing conditions (Peters *et al.* 1989a).

Several companies are developing resist materials for X-ray lithography (Peters *et al.* 1989b). Further enhancements to these resists should improve their sensitivity even further, as shown in Table 5.6 (Peters *et al.* 1989a).

Sensitivities are a function of the type of X-ray source and the wavelength. The Hunt HPR204 has a sensitivity of 300 for 1.4 nm X-rays from the Hampshire Instruments' laser source as seen in Table 5.5. For the University

Table 5.6. **Conventional resists compatible with soft X-rays.**

Resist	Present soft X-ray sensitivity (mJ/cm^2)	Optimized soft X-ray sensitivity (mJ/cm^2)
Aspect Systems		
812	200	100
AZ Hoechst		
PF–114	8	NA
PW–114	20	NA
KTI		
825	200	100
MacDermid		
914	400	150
1024	150	80
OCG Microelectronic Materials		
204	300	200
214	150	80
242	100	50
Rohm & Haas		
1029	25	NA
1054	5	NA
Shipley		
1350	250	100
1400	350	150
2400	350	150

of Wisconsin's Aladdin at 0.8 nm, the sensitivity is 2500, while a sensitivity of 1000 is measured with the BESSY source at 1.0 nm. This illustrates how the hard wavelength component from the synchrotron sources can degrade resist sensitivity.

5.4. ION BEAM SYSTEMS

5.4.1. Direct Write

There are two types of ion beam systems; focused ion beams (FIB) for direct writing on a substrate or masked ion beams (MIB) with a mask in close proximity to a resist-covered substrate.

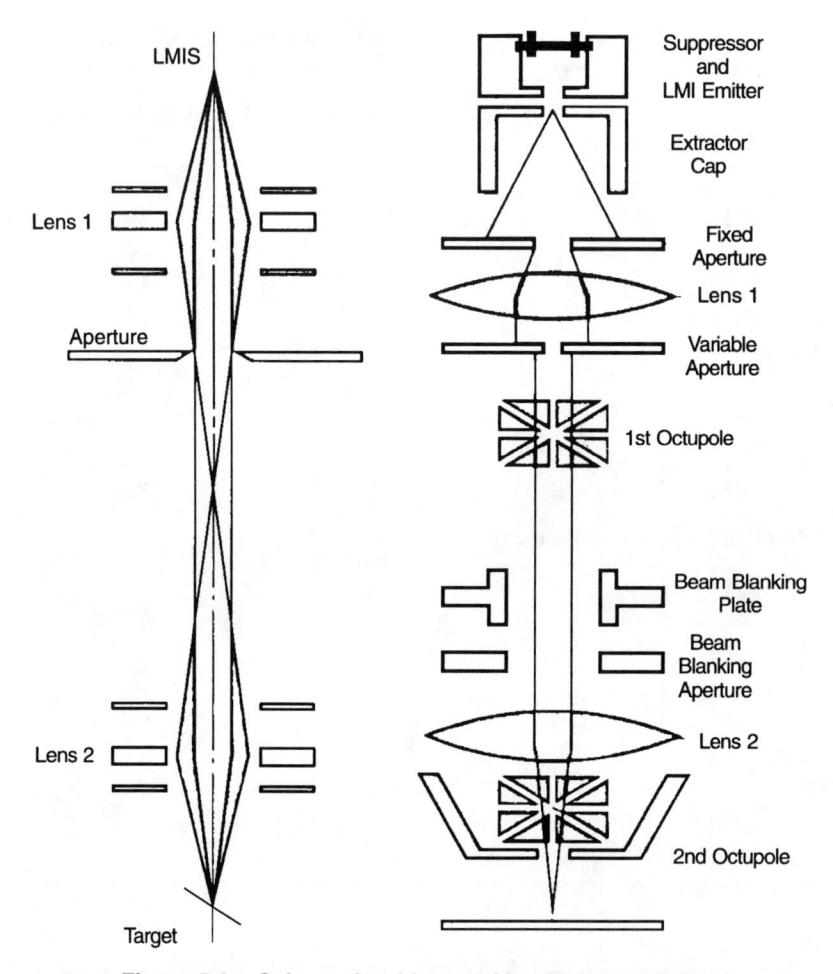

Figure 5.8. Schematic of focused ion beam systems.

In a FIB system, ions are focused on a resist-covered wafer and raster scanned across its surface, as shown in Figure 5.8. The method is similar to direct write lithography and has the advantages of higher resolution and higher registration accuracy over MIB. FIB also overcomes the disadvantages of E-beam such as secondary-electron scattering and proximity effects.

The most important physical characteristics of the FIB system are:

- Spot size (probe diameter)
- Spot current

- Deflection speed
- Field size
- Writing speed

5.4.2. Ion Channel Masking

MIB resembles full exposure X-ray systems but has the added advantage of small (0.1 µm) spot size in a step-and-repeat fashion. The image reduction makes it easier to fabricate a reticle, as with an optical stepper.

An ion channeling mask scheme for MIB has been developed at Hughes (Bartelt 1986) and is illustrated in Figure 5.9. Typically, a gold absorber is defined on a 3–6 µm-thick <100> Si membrane. High energy protons or helium (300–2000 keV) directed at the mask are absorbed by the gold and channel through the Si onto the resist covered wafer. Heavy ions, however, are inappropriate for channeling and membrane masks, due to their inability to pass through the mask material without causing damage.

5.4.3. Ion Projection

Ion projection systems for lithography (IPL) applications are analogous to optical projection lithographic systems and use a low energy ion beam (5 keV) that is accelerated to high energy (up to 400 keV, depending on system design) after passing through a self-supporting metal foil mask. The beam is then demagnified by a factor of five or ten and projected onto the substrate. Ion projection steppers operate like optical step-and-repeat systems, in which exposure is performed utilizing an X-Y stage.

The ion projection method uses an open stencil mask, with the openings evenly distributed over an area less than 50% of the mask foil material (Stengle et al. 1986). This foil is typically electroplated Ni on a bismuth and germanium doped Si substrate with thicknesses between 2 and 5 µm. The foil is mounted to a 1 mm thick Si frame and kept planar by a controlled heating of the mask frame. However, the mask fabrication process is complex. A thin silicon dioxide or nitride film is deposited over the ring area of the Si wafer. A stress-free Cr/Cu adhesion layer is then deposited over the wafer followed by the resist patterned over the wafer with a three-level resist process using oxygen and reactive ion etching. Ni is then electroplated over the wafer. The Si wafer is immersed in KOH, which etches the wafer in areas not covered by resist and Ni on top, and on the bottom ring area by the silicon dioxide or nitride film. The resist is then selectively etched.

The ion projection technique, developed by Sacher Technik (Austria), has the mask located two meters from the wafer. IPL technology has now

118 LITHOGRAPHY

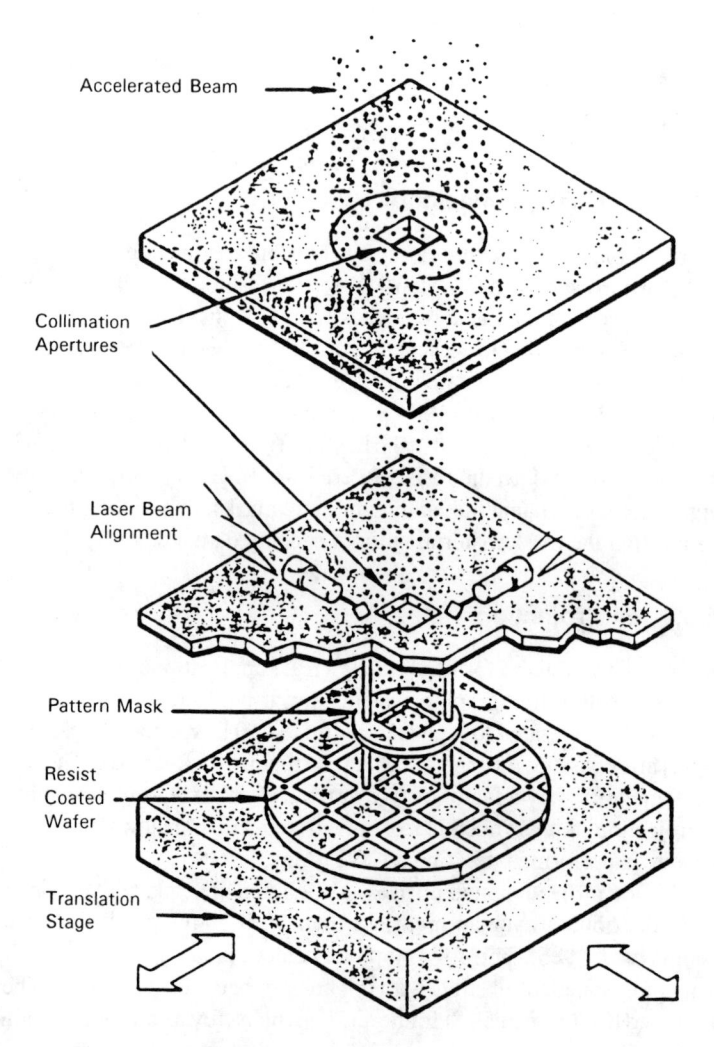

Figure 5.9. **Schematic of masked ion beam system.**

been demonstrated by Ion Microfabrication Systems GmbH (Vienna, Austria). Compared to FIB, IPL using a duoplasmatron ion source has current densities that are less than three orders of magnitude, but have 4-6 orders of magnitude shorter exposure times. Because ions impinge on the mask at low energy and are then accelerated to higher energy, the mask is subject to a less severe heating problem than with FIB.

5.4.4. Ion Sources

Liquid Metal. A FIB source consists of an ion optical column containing an ion source, extraction electrodes, mass separator, electrostatic focusing lens, and deflection system. The key element is the ion source. Several liquid metal (LM) (Clampitt, 1975) and field ionization (FI) (Levi-Setti *et al.* 1982) sources have been developed for use in lithography.

The LM source consists of a sharpened tungsten needle, 0.02 to 0.002 cm in diameter, wetted with a thin film of molten metal such as gold or gallium. Ionization of the molten metal takes place at the apex of the needle when a critical voltage (Taylor) is applied. The interaction of electrostatic and surface tension forces between the tip and exterior electrode causes the liquid metal to form a peaked cone (Taylor cone) of a small diameter. The Taylor voltage is proportional to the electrode spacing and surface tension. LM sources have a typical brightness of 1×10^6 (10E6) A/cm^2/sr (amperes per square centimeter per solid angle) at an angular current density of 10–50 microA/sr. The energy spread is 5–10 eV at these conditions, with a virtual size of 100–500 A.

Positive ions of any low melting point, low vapor pressure metal can be generated, although Au, Ga, and Li are the most common (pure materials such as Li have such high chemical reactivity that their lifetimes as an ion source is unreliable and short-lived). The energy spread has been found to increase with both current and particle mass. The energy spread for Ga, In, and Bi LM sources are 5, 14, and 21 eV, respectively, at an angular intensity of 20 microA/sr. The utilization of LM sources has been increased by the use of eutectic alloys with low melting points rather than pure elements. Alloy constituents are separated by ExB velocity filters. With heavy ions, low penetration depth and high sensitivities strongly influence the choice of resist materials for patterning.

The lifetimes of the sources are directly related to the corrosion of the tungsten needle. The reaction of the tungsten and the molten source increases the work function of the needle and results in a decrease in brightness. New studies are aimed at utilizing eutectic alloys to lower the reaction temperature as well as supplying a metal source to the needle that is noncorroding.

Field Ionization. FI sources can operate with hydrogen, argon, oxygen, or nitrogen gases and are based on the field ion microscope developed in 1951. In the FI source, a sharp tip is biased at approximately 4 keV, positive with respect to the extractor electrode. Molecular hydrogen is drawn into this field by dipole forces and becomes partially ionized and accelerated toward the liquid helium-cooled tip.

FI sources offer higher brightness, smaller energy spread, smaller ion current, and smaller virtual size than a LM source. FI sources have a typical brightness of 10E7 A/cm^2/sr at an angular current density of 10–50 microA/sr. The energy spread is 1 eV resulting in a virtual beam size of 10 Å.

Both LM and FI sources have their optimum application in VLSI technology; FI for lithography and LM for ion implantation. There is sufficient overlap, however, that LM sources can be utilized for essentially all applications of IC manufacturing including mask repair. The major disadvantage of the FIB technology is slow scanning speed resulting in low throughput. For example, with a resist sensitivity of 1E13 ions/cm^2 and a beam current density of 1 A/cm^2, writing times of 2 minutes to 1 day per cm^2 for feature sizes of 0.5–1.0 μm results. This is in part due to the need to develop sources with higher brightness and longer lifetimes. Ion source technology represents the area most critical for the technology to progress.

Investigations have shown that when Ga (or other) ions are used for machining or ion implantation applications, they are implanted into the substrate to a depth of 250 angstroms and cause lattice damage because of their high energy. This would effect the electrical properties of the device, giving rise to large leakage currents. Additional concerns are co-implanted impurities from the tungsten needle or an alloyed material, as well as redeposition of the metal over the entire wafer (Shaver and Ward 1985).

Studies at Hughes research have shown that no degradation of a shallow junction bipolar transistor was experienced when FIB was used to implant As ions. However, initial results did indicate that diode quality and junction leakage appeared somewhat degraded (Murray 1986). Ga can be removed by immersing the wafer in HF or NaOH, or in situ in Seiko's system.

FIB systems will benefit from the increased use of ASICs that require a larger number of mask sets than jelly bean devices. Also, difficulties that merchant mask makers are encountering in the competitive mask making market will force captive mask making facilities to consider the purchase of FIB systems for mask repair described below. At least 95% of E-beam made masks require repair. Several vendors practice 100% repair. Only 75% of the optically produced masks (pattern generators and photorepeaters) require repair.

The focused ion beam market will grow in two stages. The market entry will be in the mask repair application area. This method has proven advantages compared to laser repair and mask repair does not require high throughput. The second stage of market acceptability will occur when technological problems such as low throughput are resolved.

Although FIB has been the exclusive domain of mask repair, it is now entering into the circuit repair arena. FIB suppliers have recognized the limited potential of this market and have introduced modified versions of their

LITHOGRAPHY

mask repair systems for use in IC modification. Micrion, which markets a system for mask repair, has the Micrion DMOD 900, an automatic system to modify ICs directly on the wafer. The system is priced between $700,000 and $900,000. Micrion plans to start a rental service at $400 per hour and to team with test centers across the U.S. Seiko has two SM18×00 models, which vary by wafer size and cost between $600,000 and $800,000. The DMOD incorporates an electron flood gun to neutralize charge buildup on the wafer, a concept lacking in Seiko's systems.

For use in IC modification, FIB can remove and deposit material on a circuit in a similar manner to mask repair; minor changes in a chip design could be made and tested, instead of changing masks and running new silicon. FIB can microsection ICs by milling away areas of the layered structure with the high energy ions. The area of interest is observed on the CRT of the FIB system by SIM, as discussed above. Material contrast is much greater with SIM than with SEM, and there is no special sample preparation needed.

5.5. LASER LITHOGRAPHY

Ateq (now Etec) began offering a laser direct write system using an argon laser beam at 363.8 nm in early 1990. The Waferwriter 6000, priced at $2.7 million, is targeted toward prototype and process development applications, and becomes a maskless optical alternative to optical, E-beam, X-ray, and FIB.

The Waferwriter 6000 has a maximum resolution of 0.5 μm on 3-inch to 8-inch wafers by utilizing a 0.60 NA and 20-to-1 post-scan lens combined with a 363.8-nm Ar ion laser. An automatic alignment subsystem provides an overlay registration of 0.125 μm. The system also has user-selectable bright- and dark-field alignment.

The machine can be converted readily to write masks and reticles.

Table 5.7 is a summary of the lithography methods described in this section (Heuberger 1986).

5.6. NEW TECHNOLOGIES

Two recent innovations in technology have the potential capacity to be used as lithographic tools:

- Holograms have recently been utilized by Insystems as a method for inspecting masks and wafers. The founder of Insystems has now adapted the technology to lithography and has formed Holtronic Technologies. Research was jointly conducted by the Institute of Microtechnology, Neufchatel, Switzerland.

Table 5.7. Comparison of lithographic techniques.

Parameter	Optical stepper	X-ray stepper	Electron projection	Demagnifying ion projection
Resolution (μm)	0.6	0.2	0.2–0.5	<0.1
Image filed (sq mm)	15×15	50×50	50×50	4×4
Pixels/sec	5E8	5E10	1E10	1E9
reduction scale	5:1	1:1	1:1	10:1
Mask	simple	complex	none	complex
Throughput	high	high	low	low
Depth of focus (μm)	<1	50–100	20	100
Proximity effects	yes	no	yes	no
Radiation damage	no	low	high	?

LITHOGRAPHY

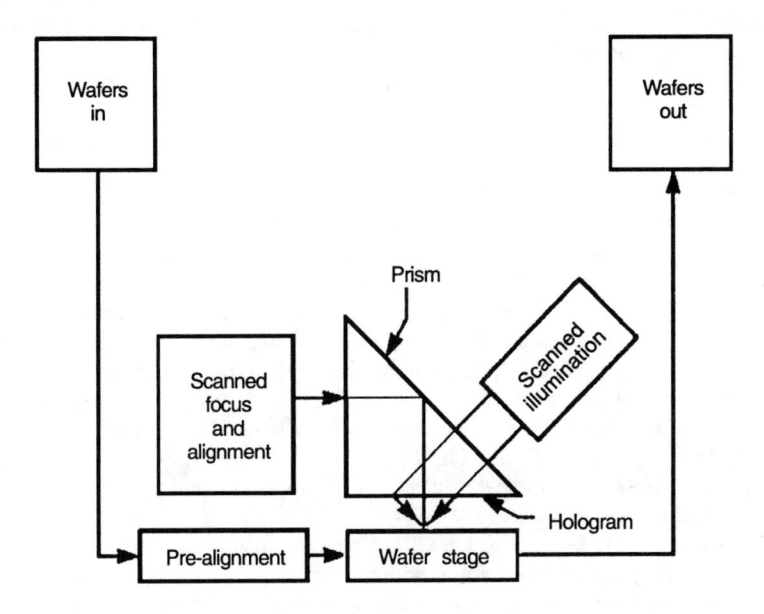

Figure 5.10. Schematic of holographic lithography system.

Resolution features of 0.3 μm have been obtained in photoresist at a wavelength of 364 nm, giving an effective NA of 0.7. Because of the close proximity between the object and imaging field, a high NA imaging system is possible, resulting in submicron resolution over an image field of 5 cm × 5 cm (Brook and Dandliker 1989). A schematic of the system is illustrated in Figure 5.10 (Omar *et al.* 1991).

- Experimental X-ray lasers produce wavelengths as short as 5 nm and are rapidly approaching 2 nm. Soft X-ray lasers have advantages of short wavelengths, coherence, and extreme brightness. This technology could rival compact synchrotron radiation for sub 0.5-μm feature sizes.

REFERENCES

Bartelt, J.L., 1986: "Masked ion beam lithography: An emerging technology," *Solid State Technology* **29** (5): 215–220.

Berglund, C.N., 1989: "E-beam direct write for IC fabrication," *Microelectronic Manufacturing and Testing* **12** (6): 1.

Brook, J. and R. Dandliker, 1989: "Submicron holographic photolithography," *Solid State Technology* **32** (11): 91–94.

Buckley, J. and Karatzas, G., 1989: "Stepper & scan: A systems overview of a new lithography tool," SPIE Proceedings, Vol. 1088.

Burggraaf, P., 1989: "Will scanner lightning strike again at 0.5 µm?," *Semiconductor International* **12** (6): 17.

Clampitt, R., K.L. Aitken, and D.K. Jefferies, 1975: "Intense field emission source of liquid metals," *J. Vac. Sci. Technol.* **12**: 1208–1211.

Dunn, P., 1990: "Bell Labs, GCA optic division build X-ray system with 20:1 reduction," *Electronic News*, April 16, pp. 28.

Heuberger, A., 1986: "X-ray lithography," *Solid State Technology* **29** (2): 93–101.

Huber, H., U. Scheunemann, E. Cullmann, and W. Rohrmoser, 1990: "Application of X-ray steppers using optical alignment," *Solid State Technology* **33** (6): 59–62.

Ittner, G., 1977: "Future possibilities of dioptric lenses in microelectronics," Proc. SPIE Conference on Optical Microlithography I, Vol. 100, pp. 115–124.

Jewell, T.E., J.H. Bennewitz, G.C. Escher, and V. Pol, 1987: "Effects of laser characteristics on the performance of a deep UV projection system," Lasers in Microlithography, Proc. SPIE, Vol. 774, pp. 124.

King, M.C., 1979: "Future developments for 1:1 projection photolithography," *IEEE Trans. Electron Devices* **ED-26** (4): 711.

Levi-Setti, R., T.R. Fox, and K. Lam, 1982: "Ion-channelling effects in scanning microscopy and ion beam writing with 60 keV Ga probe," Proc. SPIE Conference on Electron Beam, X-ray, and Ion Beam Technologies for Submicron Lithographies I, Vol. 333, pp. 158–162.

Lin, B.J., 1988: "The paths to subhalf-micrometer optical lithography," SPIE Proc. on Optical/Laser Microlithography, Vol. 922, pp. 257.

Lingnau, J., R. Dammel, and J. Theis, 1989: "Recent trends in X-ray resists: Part 1," *Solid State Technology* **32** (9): 105–112.

Malmberg, P.R., T.W. O'Keefe, M.M. Sopira, and M.W. Levi, 1973: "LSI pattern generation and replication by electron beams," *J. Vac. Sci. Technol.* **10** (6): 1025–1027.

Markle, D.A., 1974: "A new projection printer," *Solid State Technology* **17** (6): 50–53.

Markle, D.A., 1986: "Submicron 1:1 optical lithography," *Semiconductor International* **9** (5): 137–142.

Mauger, P.E. and A.R. Shimkunas, 1986: "Additive X-ray mask patterning," *Semiconductor International* **9** (3): 70–74.

Maydan, D., G.A. Coquin, J.R. Maldonado, S. Somekh, D.Y. Lou, and G.N. Taylor, 1975: "High speed replication of submicron features on large areas by X-ray lithography," *IEEE Trans. Electron Devices* **ED-22** (7): 429.

Mayer, H.E. and E.W. Loebach, 1980: "A new step-and-repeat aligner for very large-scale integration (VLSI) production," Proc. SPIE, Microlithography V, Vol. 221, pp. 9.

McCleary, R., 1988: "Performance of a KrF excimer laser stepper," Optical/Laser Microlithography, Proc. SPIE, Vol. 922, pp. 396.

McCoy, J.H., W. Lee, and G.L. Varnell, 1989: "Optical lithography requirements in the early 1990s," *Solid State Technology* **32** (3): 87–92.

Murray, C., 1986: "Functional bipolar transistors fabricated with focused ion beams," *Semiconductor International* **9** (4): 30–32.

Nagel, D.J., R.R. Whitlock, J.R. Greig, R.E. Pechacek, and M.C. Peckeran, 1978: "Laser plasma source for pulsed X-ray lithography," Proc. SPIE, Development in Semiconductor Micro-Lithography III, Vol. 135, pp. 46.

Omar, B.A., F. Clube, M. Hamidi, D. Struchen, and Simon Gray, 1991: "Advances in holographic lithography," *Solid State Technology* **34** (9): 89–94.

Peters, D.W., J.P. Drumheller, and R.D. Frankel, 1988: " Application and analysis of production suitability of a laser-baser plasma X-ray stepper," Proc. of SPIE, Vol. 923, pp 28–35.

Peters, D., D. Tomes, S. Preston, and R. Grant, 1989a: "Use Novolak-based resists with soft X-ray exposure," *Semiconductor International* **12** (5): 156–161.

Peters, D.W. and R.D. Frankel, 1989b: "X-Ray lithography: The promise of the past and reality of the present," *Solid State Technology* **32** (3): 77–81.

Plotnik, I., R.D. Frankel, and D.W. Peters, 1987: "Engineering of reticles for laser-based X-ray plasma sources," *Microelectronic Manufacturing and Testing* **12** (11): 8–10.

Preston, S., D.W. Peters, and D.N. Tomes, 1989: "Electron-beam, X-ray, and ino-beam technology: Submicrometer lithographies VIII," Proc. of SPIE, Vol. 1089, pp 166–168.

Resor, G.L. and A.C. Tobey, 1979: "The role of direct step-on-the-wafer in microlithography strategy for the '80s," *Solid State Technology* **22** (8): 101.

Rice, S. and K. Jain, 1984: "Reciprocity behavior of photoresists in excimer laser lithography," *IEEE Trans. El. Devf.* **ED-31** (1) 1–3.

Shaver, D.C. and B.W. Ward, 1985: "Semiconductor applications of focused ion beam micromachining," *Solid State Technology* **28** (12): 73–78.

Smith, H.I., D.L. Spears, and S.E. Bernacki, 1973: "X-ray lithography: A complementary technique to electron beam lithography," *J. Vac. Sci. Technol.* **10** (6): 913.

Spiller, E. and R. Feder, 1977: "X-ray lithography," *Top. Appl. Phys.* **22**: 35.

Stengl, G., H. Loschner, and J.J. Muray, 1986: "Ion projection lithography," *Solid State Technology* **29** (2): 119–126.

Thomson, M.G.R., R. Liu, R.J. Collier, H.T. Carroll, E.T. Doherty, and R.G. Murray, 1987: "The EBES4 electron beam column," *J. Vac. Sci. Technol. B* **5** (1): 53–56.

Voisin, R., 1990: "1X reticle manufacturing advantages," *Semiconductor International* **13** (6): 102–107.

Wilson, A.D., 1986: "X-ray lithography: Can it be justified?," *Solid State Technology* **29** (5): 249–255.

Yokoyama, H., F. Uesugi, S. Kishida, and K. Washio, 1985: *Appl. Phys.* **A-37**: 25.

Chapter 6

MASK MAKING, INSPECTION AND REPAIR

6.1. MASK MAKING

6.1.1. Mask Blanks

The choice of photomask material depends on the device geometry and tools and techniques used to make it. With the increased use of projection and stepper aligners, and the need for high registration, there is a trend toward the use of materials with low coefficients of expansion. As a result, there was a shift from soda lime to low expansion glass in the early 1980s to quartz in the mid-1980s. In the U.S., soda lime blanks are used almost exclusively for contact/proximity aligners, which have been supplanted by alternative optical and non-optical methods of pattern delineation.

The Japanese, however, have recognized the advantages of quartz and use this mask material even on contact/proximity aligners. Most of the MOS memory product masks made in Japan have been made with quartz substrates

128 MASK MAKING, INSPECTION AND REPAIR

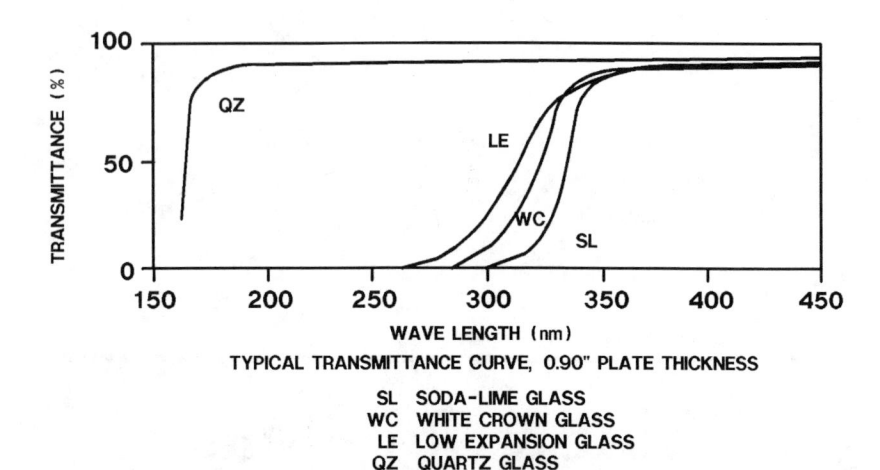

Figure 6.1. Light transmittance of glasses.

TYPICAL TRANSMITTANCE CURVE, 0.90" PLATE THICKNESS

SL SODA-LIME GLASS
WC WHITE CROWN GLASS
LE LOW EXPANSION GLASS
QZ QUARTZ GLASS

since 1984. This occurs despite the fact that approximately 50% of the installed base of alignment equipment is either of the contact or proximity type. Quartz is growing strongly because of the low and consistent coefficient of expansion and the flat optical transmission curve over the wavelengths most commonly used. Quartz efficiently transmits light over the entire spectrum from 180 to 400 nm, as shown in Figure 6.1. Since it is harder and more stable than soda lime or low expansion glass, quartz can be polished to a higher quality finish that reduces polishing defects. However, the adhesion of chrome or iron oxide becomes more of a problem. Synthetic quartz offers the best and most predictable consistency with respect to transmission, surface finish quality, and lack of bubbles and inclusions. Nevertheless, a variety of proprietary steps are used by blank suppliers to assure the high levels of quality of the material and the degree of coating adhesion.

The market for photomask blanks and substrates is dominated by the Japanese. For every U.S. vendor that has entered the market since the 1970s (when the Japanese entered the commercial market), five have dropped out of the business, unable or unwilling to compete on price or quality.

Glass and quartz substrates are now the exclusive domain of Japanese manufacturers. Hoya supplies soda lime and low expansion glass, and has about 70% of the worldwide chrome blank market, including a significant and growing portion that is made of quartz. The quartz substrate market is dominated by Shin-Etsu, and is shared by Toshiba Ceramics with Asahi Glass coming on strong. Shin-Etsu has about 70% of the substrate market. Hoya is

MASK MAKING, INSPECTION AND REPAIR

developing its own quartz, and there are several other minor suppliers. Indications are that some U.S. manufacturers may consider entering this market, which Corning abandoned in the early 1980s. There are indications that U.S. semiconductor manufacturers would be pleased if a U.S. vendor of both quartz and chrome blanks would become a force in this marketplace. The leading blank vendors are:

- Dai Nippon Printing
- Hoya Electronics Co
- Toppan

Other mask blank vendors are:

- Balzers—Europe and specialty products in U.S.
- DuPont—formerly Tau Labs
- Inco—minor
- Telic—minor
- Ultraglass—minor
- Ulvac—Japan

6.1.2. Completed Masks

The starting point in mask or reticle fabrication is a defined image or artwork of a die at one mask level. If this pattern is in the form of a drawing, it has to be digitized and stored on a magnetic tape. A typical photomask fabrication flow is shown in Figure 6.2.

A pattern generator, as used here, means the device used to convert the design data to a physical pattern on a mask plate. The product of the pattern generator is usually a reticle, which in mask making is nominally 10X. However, mask makers fabricate reticles for use in wafer steppers, which may be anywhere from 1X to 10X. The optical pattern generator is a vector plotter, in that the table is vectored to the point where a shape, usually rectangular, is exposed with a flash.

The critical layers of an LSI or VLSI pattern requires a prohibitively high flash count. As a result, the optical pattern generator takes too long and is uneconomical for these applications. Most E-beam machines convert the vector data to raster form, and are largely unaffected by pattern complexity, as long as they meet the basic precision required.

If a critical mask (1X) is to be fabricated optically, the reticle image (10X or 10 times the size of the die, which is then reduced) is stepped repeat-

130 **MASK MAKING, INSPECTION AND REPAIR**

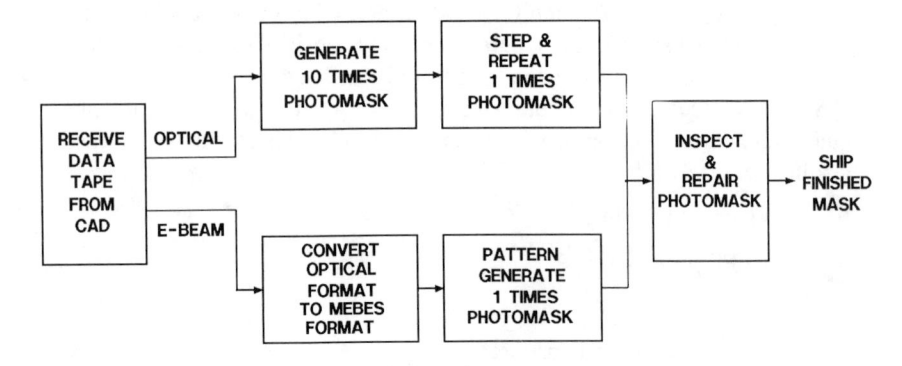

Figure 6.2. Photomask fabrication flow.

edly across a chrome blank, thus creating a 1X mask. In contrast, an E-beam machine makes a 1X mask by rastering through the whole plate directly, being driven from the digitized design date without an intermediate reticle step. Most E-beam generated plates are reticles of various magnification used either in a wafer stepper or a mask stepper. Photomask fabrication steps for optical and E-beam are illustrated in Figure 6.3 and Figure 6.4.

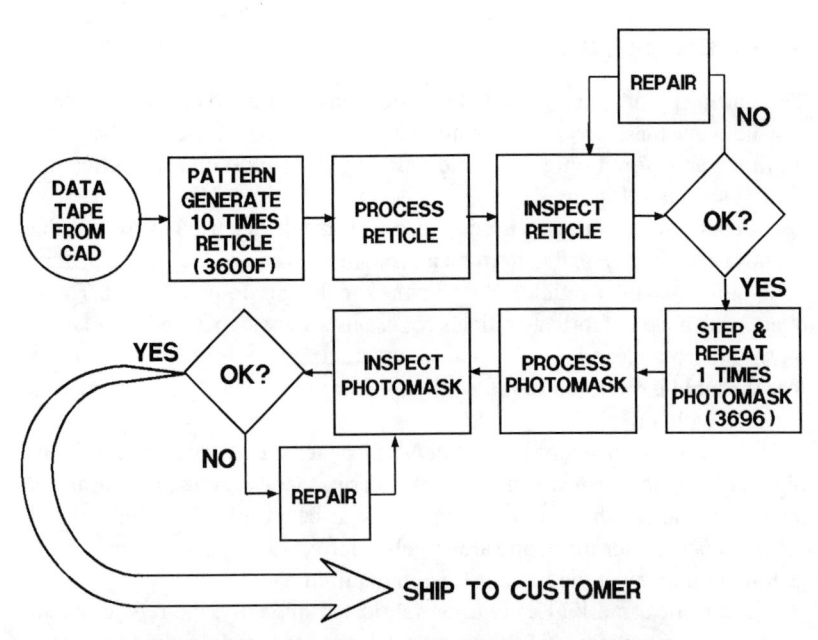

Figure 6.3. Optical photomask fabrication flow.

MASK MAKING, INSPECTION AND REPAIR

Each of the eight largest U.S. mask makers listed below had 1989 revenues of $5–20 million and has at least one E-beam system:

- Align-Rite
- Master Images, acquired by Du Pont (1986)
- Microfab, acquired by Photronic Labs (1988)
- Micro Mask, acquired by Hoya (1989)
- Photronic Labs, acquired Sierracin (1986)
- Qualitron, acquired by P-E
- Tau Labs, acquired by Du Pont (1985)
- Rexotech, acquired by Du Pont (1989)

Other Merchant U.S. companies include:

- Adv. Reproductions
- Diamon Images
- Micro Phase
- Perkin–Elmer, sold to DuPont (1989)
- Photo Sciences, all optical
- Tektronix, sold to DuPont (1991)

Foreign companies include:

- Align-Rite, Wales
- DNP, Japan #1 in world, also markets blanks

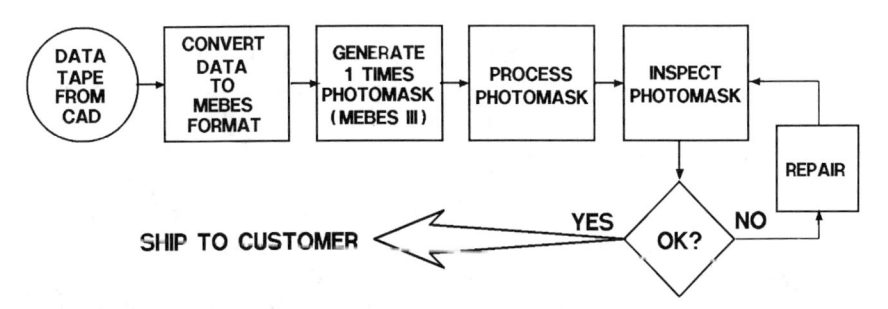

Figure 6.4. E-beam photomask fabrication flow.

132   MASK MAKING, INSPECTION AND REPAIR

- Hoya, Japan #1 in world in blanks
- Compugraphics, UK acquired by Laporte
- IC Mask, UK
- Innova, Taiwan
- Nanomask, France, purchased by Du Pont
- Taiwan Photomask, Taiwan
- Toppan, #2 in world, bought TI's mask shop

The business for masks for new layers seems to be more steady than that of semiconductors or for mask reprints, since in a recession the number of new designs increases, but the number of wafers made does not. The IC average sale price starts a new learning curve upon the beginning of smaller production lots of wafers, but the mask makers have to suffer price erosion. Since approximately three-quarters of all masks made are new designs, mask making is a more stable business than semiconductor fabrication. This stability is reflected in the unit volume of mask plates and leads to more stability for the mask business than any other equipment.

Responding to the needs of their customers, merchant mask makers are becoming much more reliable. In fact, the trend is toward closer cooperation and collaboration between the merchants and their customers. There is evidence that agreements between merchants and their customers are being patterned after the oil industry involving a "take or buy" arrangement. It is probable that Japanese mask vendors enjoy the benefits of similar deals. These may not be formalized, but they nevertheless account for a very large share of the market the Japanese merchants command (60%).

A recent advancement in mask technology is the use of phase-shifting masks. Phase-shifting masks can improve stepper resolution up to 100%. Eventually, phase shifting masks may be used with deep-UV steppers to make structures with critical dimensions down to 0.15 µm (Lin, 1990).

A combination of phase shifting masks, still in early stages of development, 0.48 NA lenses, and high contrast resists will make it possible to extend I-line equipment for 16Mbit DRAMs for mask levels with 0.5-µm feature sizes and some levels of 64Mbits. Eventual utilization with excimer-laser based optical systems could be used on the 1Gbit DRAM in the year 2001.

Although phase-shift masks currently cost ten times that of a standard optical mask, its development and success could be crucial for future technology. The window for purchases for 64 Mbit DRAM production (or equivalent feature sizes and complexity) is now open, even though full production will not begin until 1995. The year 1991 saw the purchase of much of the front-

MASK MAKING, INSPECTION AND REPAIR 133

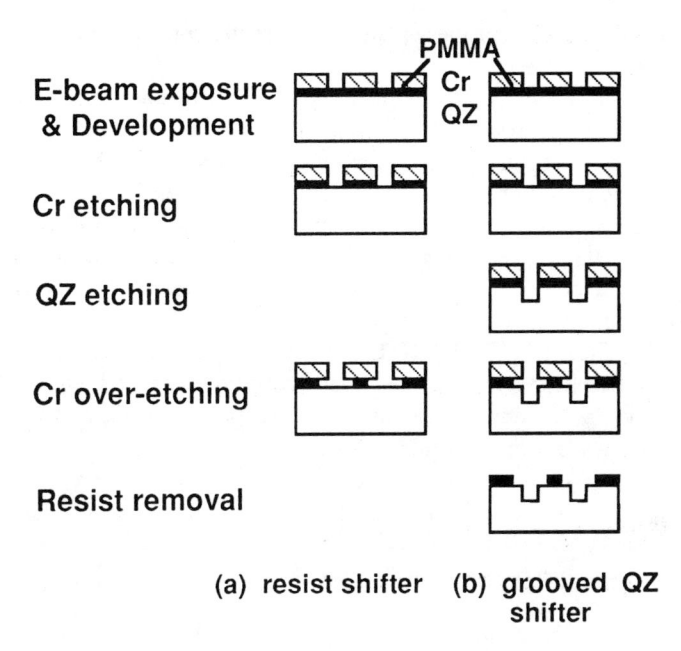

Figure 6.5. Phase shifting masks.

end equipment used to make them. Therefore, advancements must be made quickly, before alternative types of lithography equipment are utilized because phase-shifting is not in place.

Among the earliest results of phase-shifting masks was reported by engineers of Hitachi's Central Research Laboratory, extending the resolution of an I-line stepper to 0.3 μm (Burggraaf 1989). Engineers at Matsushita Electronics' Kyoto Research Laboratory have announced that their "self-aligned phase-shift masks with a resist shifter" is suitable for fabricating contact holes. The fabrication steps for two versions of the mask are shown in Figure 6.5 (Todokoro *et al.* 1991). In one version (a), the electron beam exposed patterns serve for mask chrome etching and then remain as phase shifters. In version (b), additional fabrication steps produce grooved-quartz shifters that are more stable than resist shifters. In addition to fabrication simplicity, these two processes provide a compactness of mask data compared to other phase-shifting masks.

At the University of California at Berkeley, graduate students are characterizing projection printing with phase-shifting masks based on simulation of aerial images with SPLAT and simulation of resist development with SAMPLE, to provide a greater understanding of phase-shifted image forma-

134 MASK MAKING, INSPECTION AND REPAIR

tion and its dependence on mask pattern layout and stepper parameters (Neureuther 1990).

Engineers at Fujitsu are generating phase shift masks using a self-aligned fabrication process (Nakagawa 1991). Chrome patterns of 0.3 µm and 0.15 µm shifter patterns are formed simultaneously by isotropic etching.

At Hitachi's Central Research Laboratory, engineers have developed a subshifter layer that aids in phase shift mask repair (Okazaki 1991). With a defect in the shifter layer, they remove both the shifter and the subshifter layers around the defect, making the phase at the repaired site 360° different from the original phase condition.

At Oki Electric's VLSI R&D Center, engineers have developed a phase shift mask method that enables them to focus different device features in different planes, permitting them to tailor the focal position of each feature to a particular surface plane (Ohtsuka 1991).

6.2. MASK MAKING EQUIPMENT

6.2.1. Optical Pattern Generators

Optical pattern generators and photorepeaters are built by GCA while E-beam mask generators are becoming a near monopoly of Perkin-Elmer (now Etec, purchased in 1990 by IBM, Grumman, DuPont, Micron Technology, Zitel, and a group of EBT division managers). Others supplying special E-beam equipment for this purpose are Lepton, Cambridge Instruments (Leica), JEOL, Toshiba, Nikon, and Hitachi. Laser pattern generators have been introduced by GCA, and ATEQ. The active installed base of optical pattern generators is estimated to be around 550, while that of E-beam mask making machines is about 150.

6.2.2. Electron Beam Systems

As described above, the shorter wavelengths of the illumination source has extended the resolution of optical systems to the 0.35 µm linewidth region. The smaller wavelength of electrons compared to optical (visible and ultraviolet) radiation permits resolution as small as 0.125 µm. In addition, the charged electrons can be easily vectored by electrostatic and magnetic fields facilitating faster movement of the electron beam as well as increased flexibility in writing a variety of patterns on a mask, reticle, or wafer. Optical photons, on the otherhand, must be mechanically moved by mirrors or by movement of the stage.

The most important issue impacting mask making E-beam systems is registration accuracy. Since there are no registration marks on masks, as op-

MASK MAKING, INSPECTION AND REPAIR

posed to wafers, the referencing must be within the system itself. This places stringent requirements on the mechanical stability of the E-beam system. A single layer written across a 6-inch mask requires a high degree of stability to maintain registration accuracy as no interim realignment is performed. In order to achieve a 0.1 µm overlay accuracy, precise temperature control and table positioning are required. A one degree centigrade change in temperature can cause misregistration in two successive glass mask levels as large as 0.25 µm over a 3-inch area.

Sophisticated E-beam systems, selling for as much as $4 million, utilize the computer-aided-design (CAD) pattern output to generate the masks and reticles. Software capabilities of the E-beam computer are used to control writing functions such as stage motion, beam deflection, fiducial mark detection, and blanking speed.

Nevertheless, errors in mask fabrication occur because of several factors:

- Incorrect E-beam parameters
- Errors in syntax
- Wrong titles
- Misplaced patterns
- Masks that are not optimized for manufacture

AT&T Bell Laboratories' MASKVIEW software has been used since 1984 and was commercially licensed MASKVIEW in 1987. MASKVIEW offers the following features:

- Displays mask data
- Simulates the mask
  — Displays proper pattern placement
  — Displays proper titles
  — Displays proper tone
  — Displays proper feature size
- Verifies manufacturability
- Simplifies data analysis
  — Summarizes and identifies
  — Measures and overlays
  — Zooms and pans
  — Displays boundaries
- Improves data communication

136 MASK MAKING, INSPECTION AND REPAIR

A problem with E-beam-generated masks and reticles is roughness of mask edges, particularly for 1X masks. This has been reduced in recent years by improved electronics. A significant improvement has been attained by generating a 10X reticle with an E-beam and then using a conventional step-and-repeat camera to produce the 1X mask.

The major technological advantage of E-beam mask-making systems compared to optical techniques is its increased speed and better accuracy. Complex circuits that would require over 10 hours to generate on an optical pattern generator can be made in less than an hour with an E-beam system. In addition, overlay accuracy of 0.1 µm can be attained with an E-beam compared to 0.4 µm with an optical pattern generator. Edge roughness has become less of a problem with the replacement of raster scan technology by vector scan technology.

In addition to the 0.1 µm overlay accuracy, E-beam generated 1X masks are commonly fabricated with defects of 0.5–0.9 µm, with CD (critical dimension) tolerances of 0.1–0.15 µm. For devices with 0.75 to 1.0-µm feature sizes, overlay accuracies of 0.05–0.1 µm, defects of 0.25–0.5 µm in size (defect density of 2.5/in.^{2}), and CDs of 0.05–0.1 µm have been achieved. For the 64 Mit DRAM, requirements will be overlay accuracies of 0.03–0.04 µm, defects of 0.1–0.25 µm, and CDs of 0.025–0.03 µm for devices fabricated with 0.4-µm feature sizes.

6.2.3. Laser Pattern Generators

Laser pattern generators have been introduced that also use raster methods. They can make 5X and 10X reticles more economically than E-beam equipment as well as 1X reticles or masks with newer software capabilities. Another advantage is that they can use standard optical resists, which are easier to work with and which create fewer and smaller defects of 0.6 to 0.9 µm than E-beam resists.

The Japanese were the first to recognize their potential and received the first two Core-2000 systems from Ateq, illustrated in Figure 6.6. Installations to U.S. manufacturers have followed. Ateq's Core-2500 is priced at $2.5 million. While the Ateq system uses an Ar laser (363.8 nm) (Burns and Schoeffel 1987), Micronic Laser Systems (Täby, Sweden) uses a HeCd laser (441.6 nm) (Moretti, 1991).

The advocates of laser pattern generators suggest that they will find applications in direct write on wafers as well. The customers, wafer fabs and particularly the ASIC manufacturers, would like to take advantage of the faster time and fewer defects that this promising technology affords.

MASK MAKING, INSPECTION AND REPAIR 137

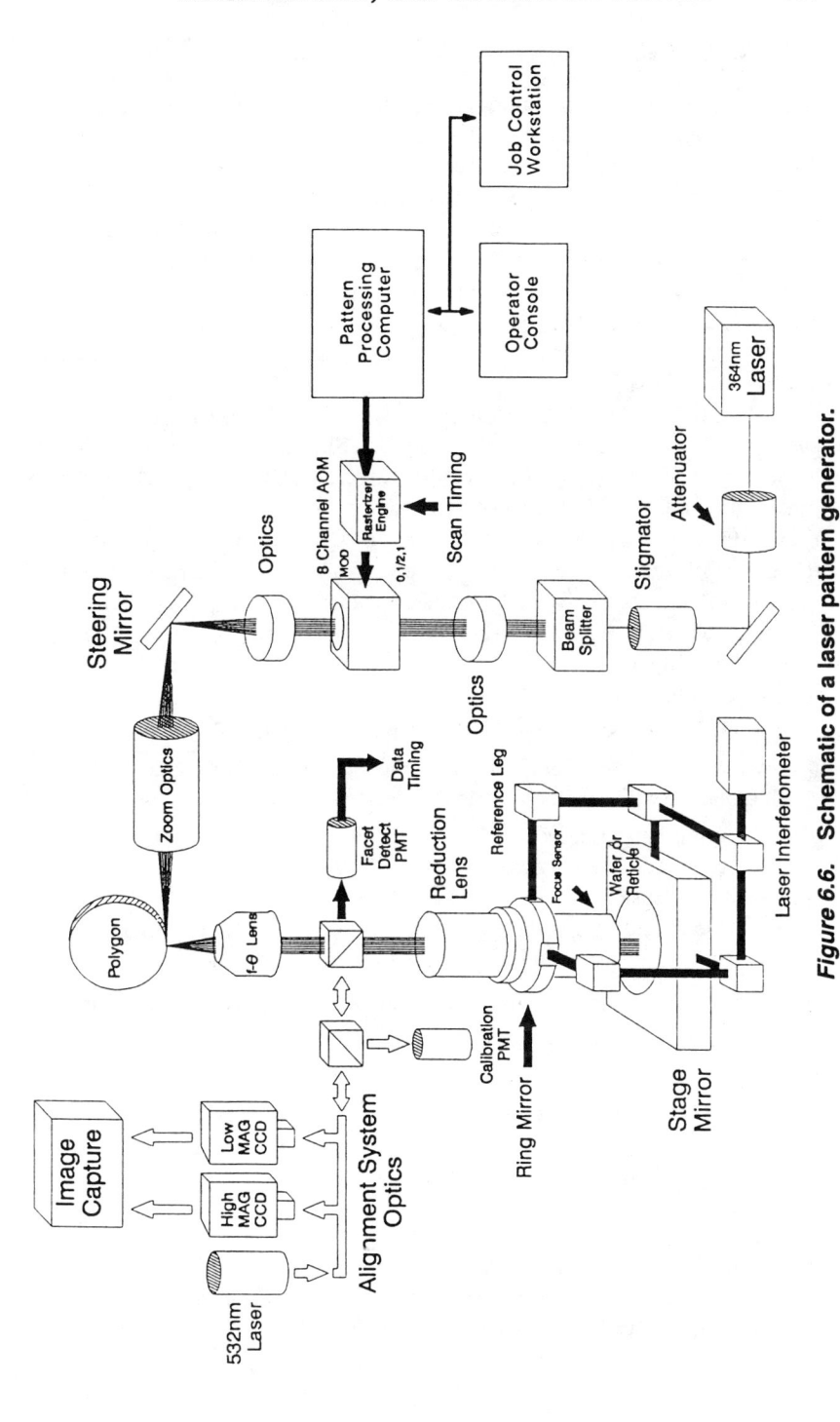

Figure 6.6. Schematic of a laser pattern generator.

138 MASK MAKING, INSPECTION AND REPAIR

6.3. MASK INSPECTION

In the early 1980s, it was possible to adequately inspect masks and reticles by visual inspection. As linewidths have decreased, it is estimated that 50% of defects below 2 μm in size would be missed by a visual inspection.

Reticles must be defect free, as any defect would be repeated on each die image, whether on a mask or on a stepped wafer. Even on directly-generated 1X E-beam masks, defect density must kept to the barest minimum. The term error budget, borrowed from programming language, takes on added significance when applied to 0.25 μm-size defects. Consequently, 100% defect inspection of critical inspection of critical layers will be required as long as masks are fabricated.

There are two major concerns of mask makers who use automated inspection systems—sensitivity and throughput. Shown in Figure 6.7 is an illustration of die-to-die and die-to-database inspection. State-of-the-art sensitivity is currently 0.2 μm, capable with the KLA 331. Sensitivity, however, is the basic need until the required minimum level needed for a device is reached. At this point, throughput becomes important. The KLA 331 can inspect a 100-mm^2 reticle for 0.20 μm defects in 60 minutes, a 0.40 μm defects in 15 minutes, and 0.60 μm defects in 7 minutes. By contrast, the fastest previous KLA system, the 239HR, requires 155 minutes to detect 0.40 μm defects on a 70-mm^2 reticle.

Other factors determined in the course of this study are:

- *Automation*, with specific recommendations (that also include issues associated with repair equipment described above) that include
  — Automatic align/locate/place
  — Automatic diagnostics
  — Autofocus
  — Robotic transfer from inspection to repair and to re-inspect
  — Automatic inspection and repair—an integrated system
- *Spot size* has been addressed by the new KLA 230e KLARIS Systems.

Current inspection capability can be extended to 0.15 μm detection on 0.3-μm linewidths with visible light. With UV radiation, detection can be extended to 0.1 μm on 0.2-μm lines.

6.4. MASK REPAIR

On the average, at least 95% of E-beam made masks require repair. Several companies practice 99% and 100% repair, while fewer (about 75% on the average) of the optically (pattern generators and photorepeaters) produced masks require repair.

Mask Making, Inspection and Repair 139

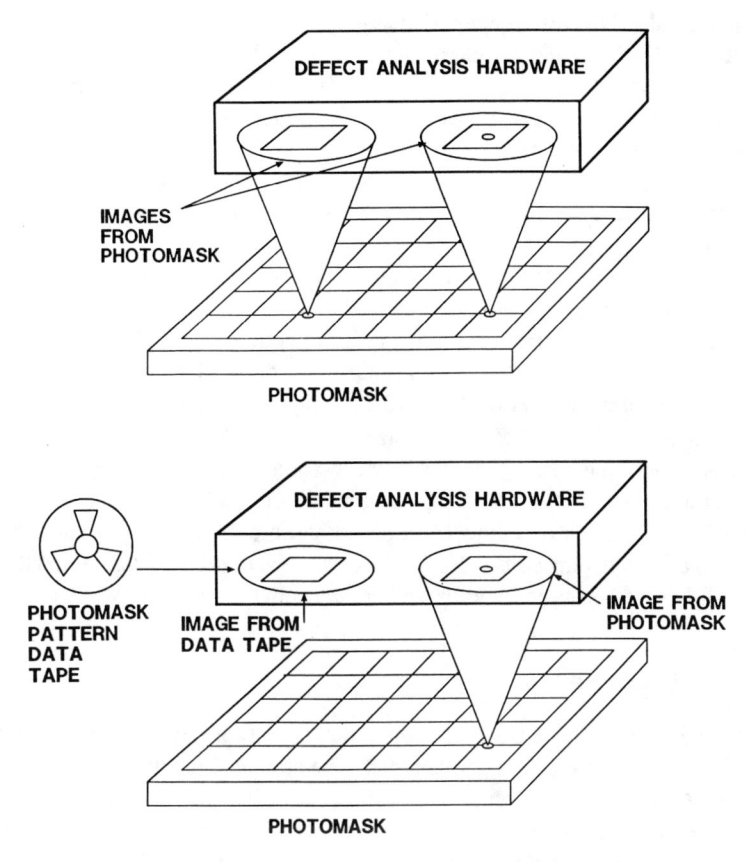

Figure 6.7. Die-to-die (top) and die-to-database (bottom) inspection.

Only a small percentage of the ones needing repair are scrapped, certainly less than 5%. The criterion to scrap is judgemental. One interesting guideline is to scrap if, in the judgement of the operator, the repair would take more than 15 minutes. This indicates a certain speed expectation for repair equipment. It can be assumed that about half of the masks scrapped are subjected to repair, or at least to a classification judgement. This means that even the scrapped ones require repair equipment and time, at least until better decision making software for inspection is developed. Of the masks needing repair, at least 95% can be repaired, and of those which are repaired around 95% are usable.

Our quantitative conclusion on repair load—i.e., the number of masks repaired as a percentage of masks made—is 97% for E-beam masks and 77%

140 MASK MAKING, INSPECTION AND REPAIR

optical masks. This difference is due largely to the characteristics of E-beam resists (PBS, COP) and optical ones.

In general, it was found that the considerations for opaque and clear defects as far as their sizes are concerned are the same. "A defect is a defect" was a refrain of those who gave consideration to both. But it was also found that several mask makers did not repair clear defects at this time. Also, E-beam resists seem to produce more clear defects than optical ones as well as smaller minimum defect sizes.

The number of defects per area that are acceptable will decrease each year. While zero defects are often mentioned, most are more realistic at the 1X level, particularly if these are for masks rather than 1X reticles. Responses from our survey show that the current acceptable defect density is 2.5 defects/cm^2, of 0.35-µm sizes. For 5X and 10X reticles the current requirement is a defect density of 0.5/in^2 of 1.5-µm size.

Another factor for 1X masks is not the number of defects per area per plate, but the number or percentage of good dies or bad dies per plate. (Who cares if one die has one thousand defects and the others are perfect?) Yet another criterion, particularly for 1X reticles, is the number of defects per die, (usually expressed as good fields) rather than with respect to the surface area.

6.4.1. Laser Repair

CW Ar ion lasers have been used for clear (defect removal) repair, while a Nd:YAG laser is used for opaque repair. The chemical reaction for clear repair can be driven by two methods; laser photolysis and laser pyrolysis. In *laser photolysis*, a laser is used to decompose an organometallic molecule such as cadmium dimethylene in the vapor phase above the clear defect. Although submicron repair is possible, this method has several disadvantages:

- Second harmonics that are generated lead to inefficiencies in the process requiring a large, high-power argon laser

- These second harmonics exhibit an output that fluctuates with time, requiring periodic adjustments

- An argon laser operating at 257 nm requires costly, high-numerical aperature, long working distance quartz optics

With *laser pyrolysis*, the chemical reaction occurs at the surface of the mask, while the volatile components remain in the vapor phase. Using this technique, clear defects have been repaired with published results approaching 1.6 µm. However, features of 1 µm have been repaired by using a combination of laser pyrolysis and laser trimming. Nevertheless, a practical limit of 1 µm is expected. This method has a number of advantages:

MASK MAKING, INSPECTION AND REPAIR   141

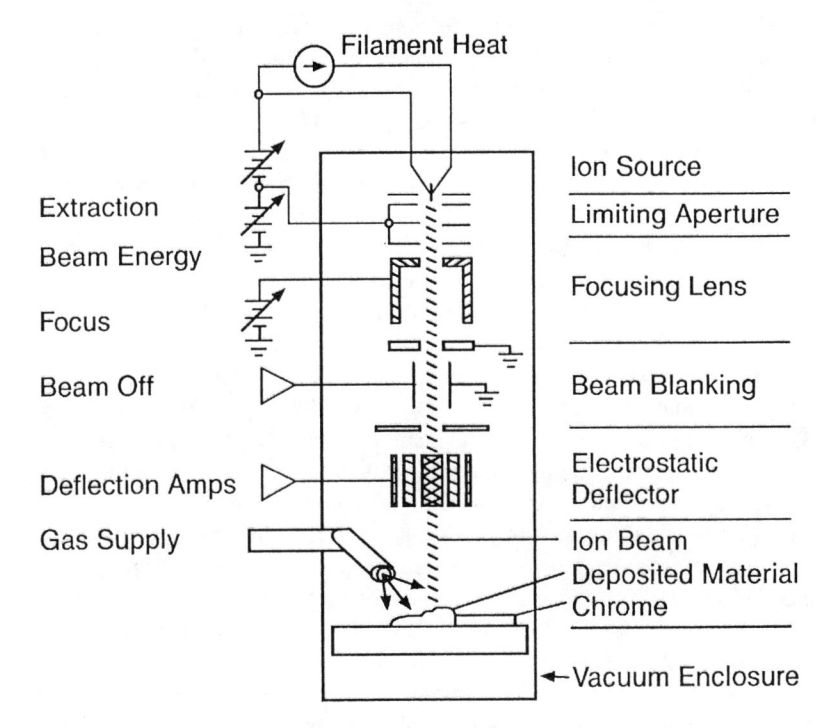

Figure 6.8. Schematic of a focused ion beam system.

- Depending of the optical delivery system, a compact 50 mW air cooled argon ion laser suffices as a source
- Argon ion laser output is tunable and stable over long periods of time
- Since the pyrolytic deposition utilizes a visible laser, no expensive UV or IR optics are required.

6.4.2. Focused Ion Beam Repair

A focused ion beam (FIB) method, illustrated in Figure 6.8, proves a viable alternative to laser repair that is limited to a 1-μm focused laser spot (because of the diffraction of light).

The FIB source for mask repair consists of an ion optical column containing an ion source, extraction electrodes, mass separator, electrostatic focusing lens, and deflection system.

The key element is the ion source. Several liquid metal (LM) sources have been developed for use in mask repair.

Positive ions of any low melting point (Culbertson *et al.* 1979), low vapor pressure metal can be generated, although Au, Ga, Si, and Li are the most common (pure materials such as Li have such high chemical reactivity that their lifetimes as an ion source is unreliable and short-lived). The energy spread has been found to increase with both current and particle mass. The energy spread for Ga, In, and Bi LM sources are 5, 14, and 21 eV, respectively, at an angular intensity of 20 microA/sr. The utilization of LM sources has been increased by the use of eutectic alloys with low melting points rather than pure elements. Alloy constituents are separated by ExB velocity filters. With heavy ions, low penetration depth and high sensitivities strongly influence the choice of resist materials for patterning.

The lifetimes of the sources are directly related to the corrosion of the tungsten needle. The reaction of the tungsten and the molten source increases the work function of the needle and results in a decrease in brightness. New studies are aimed at utilizing eutectic alloys to lower the reaction temperature as well as a metal source to the needle that is noncorroding.

However, gallium staining of the photomask occurs during mask repair. This can degrade the glass or quartz, and change the optical transmission of the photomask. Micrion and others have shown that Ga can be removed by immersing the wafer in HF or NaOH. Seiko Instruments has an accessory to their SIR-1000 that removes Ga in situ.

MicroBeam's NanoFix LM source uses no gallium, but a gold alloy. The system automatically selects the right beam to optimize each function of the mask repair process. Additional concerns are co-implanted metallic impurities from the tungsten needle or an alloyed material, as well as redeposition of the metal over the entire wafer.

Focused ion beams can be operated in the physical sputtering or ion beam assisted chemical etching modes to machine ultrafine lines. Excess metal can be removed for the repair of local defects and edge reconstruction. Clear defects in optical masks can be removed by etching the glass substrate.

Researchers at AT&T Bell Labs (Wagner 1983) have used FIB to repair both optical and X-ray masks with 0.5-μm lines and spaces at a rate of <10 sec/sq. μm. The same FIB can be also used for mask inspection, since the system can be operated as a scanning electron microscope.

There are two methods of repairing clear defects. The Micrion method repairs clear defects down to 0.25 μm by creating blocking patterns in the photomask. The substrate is machined at the defect site into a particular geometry that will act as a diffuser or reflector and prohibit the transmission of light. The Seiko and MicroBeam methods use ion assisted deposition to selectively deposit carbon and chrome, respectively, at the defect site. Figure 6.9 is an illustration of these processes.

MASK MAKING, INSPECTION AND REPAIR 143

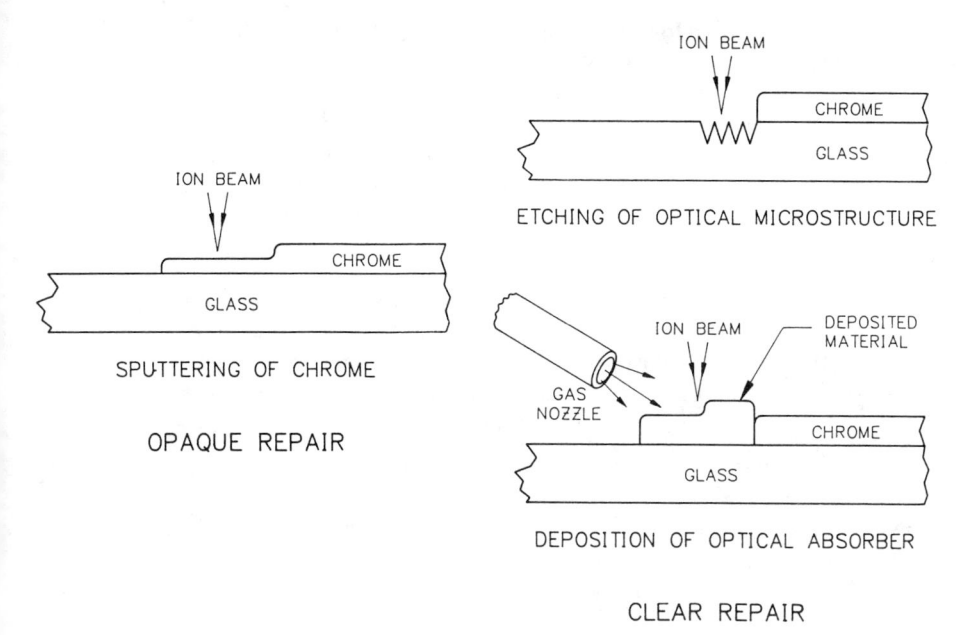

Figure 6.9. Illustration of clear and opaque mask repair.

These FIB processes compete with existing technologies such as ultrasonic cutters, lasers, mechanical probes, and deposition techniques. However, the small spot size of the FIB technique makes it a viable alternative, particularly at submicron features, as well as for X-ray, which requires Au, W, and other metals to absorb the X-rays.

Recent developments in FIB technology include:

- MicroBeam Inc. developed the NanoFab-150 that features a 3 kV to 150 kV beam. The system has a universal ion source that accomodates the latest advances in LM sources. The system is used for lithography, implantation, deposition, and selective etching.

- JEOL's focused ion beam system model JIBL-100 uses a liquid gallium source with a beam size of less than 0.1 µm at an accelerating voltage of 100 keV and an ion beam current of 100 pA. The system incorporates mass analysis of the beam species and is priced at $2 million for an advanced research machine. The systems are currently used for direct implant of prototype Si and GaAs devices and for experiments on X-ray mask making and repair. They are not currently used for mask repair.

Systems used for lithography have been sold in Japan to NEC and to the Optoelectronics Joint Research Laboratory.

- Formed early in 1984 by founders of IBT, Micrion supplies a FIB system for mask repair. The Micrion 808 system uses a liquid metal ion source and precise electrostatics to create beam diameters of approximately 0.2 µm. Micrion is currently working with Perkin-Elmer (Etec) to develop a mask repair process for MEBES E-beam machines.

- Seiko Instruments' SIR-1000 mask repair system was developed after several leading personnel left JEOL. The system offers resolution to 0.25 µm, repairing opaque defects with a continuous carbon film. Seiko entered into a U.S. distribution agreement with laser manufacturer Quantronix in 1987, but has been marketing the system directly since the agreement was dissolved in 1989.

REFERENCES

Burggraaf, P., 1989: "Extra reticle openings trick stepper," *Semiconductor International* **12** (6): 50.

Burns, G.A. and J.A. Schoeffel, 1987: "Scanning laser technology applied to high speed reticle writing," Proc. Semicon/West, San Mateo, Ca, pp. 262–277.

Culbertson, R.J., T. Sakurai, and G.H. Robertson, 1979: "Ionization of liquid metals, gallium," *J. Vac. Sci. Technol.* **16** (2): 574.

Lin, B, 1990: 10th BACUS Symposium on Microlithography, Santa Clara, CA.

Moretti, M., 1991: "HeCd Laser Writes Reticles for microlithography," *Laser Focus World* **27** (5): 29–30.

Nakagawa, K., 1991: IEEE International Electron Devices Meeting (IEDM), San Francisco, CA.

Neureuther, A., 1990: 10th BACUS Symposium on Microlithography, Santa Clara, CA.

Ohtsuka, H., 1991: IEEE International Electron Devices Meeting (IEDM), San Francisco, CA.

Okazaki, S., 1991: IEEE International Electron Devices Meeting (IEDM), San Francisco, CA.

O'Neill, T., 1981: "Photomask materials review," *Semiconductor International* **4** (9): 81–94.

Todokoro, Y., H. Watanabe, Y. Hirai, N. Nomura, and M. Inoue, 1991: "Self-aligned phase shifting mask for contact hole fabrication," *Microelectronic Engineering* **13**: 131–134.

Wagner, A. 1983: "Applications of focused ion beams," Proc. SPIE Conference on Electron Beam, X-ray, and Ion Beam Technologies for Submicron Lithographies II, Vol. 393, pp. 167–176.

Chapter 7

PLASMA ETCHING

7.1. SINGLE WAFER SYSTEMS

7.1.1. Advantages Over Batch Systems

As linewidths decrease and device densities correspondingly increase, the need for new process and equipment designs have become apparent. Presently, the plasma etching market is dominated by the batch processing system characterized with a multi-faceted, reactive ion etch reactor design. This particular design is commonly referred to as the hexode reactor. Numerous other batch reactor designs existed in the marketplace but failed to achieve the notoriety and major market share of the hexode system.

Plasma etching on a full production scale began with parallel-plate, batch-processing systems that offered the advantages of increased process control, much improved uniformities, and higher device yields over their predecessors, the barrel etchers. Evolution continued from plasma etching to reactive ion etching and variations on the theme of controlling incident ions to enhance the etch process. The hexode reactor design was born during this period of evolution, sometime in the early to mid-seventies.

Market timing for the introduction of the hex system was excellent. Plasma and reactive ion etching had come of age and were recognized as necessary in order to successfully fabricate devices that were on the leading edge of technology. Wet etching and plasma barrel etchers were incapable of taking the semiconductor industry to the next level design criteria because of severe undercutting problems. Hence the introduction, and resulting tremendous success, of the hexode reactor.

The increased complexity of devices, coupled with smaller features, has had a profound effect on both etching processes and hardware designs. The dry processing technology of reactive ion etching is emerging as the front runner for VLSI device manufacturing. User issues include requirements for wafer to wafer reproducibility, controlled wall profiles, and submicron etch resolutions.

The use of six-inch wafers was the most important issue facing equipment suppliers the past several years. Today, eight-inch wafer processing is helping to mold equipment designs. As wafer sizes increase, the cost per wafer processed is increasing accordingly. Tighter controls are required in order to maximize yield and minimize cost. As shown in Table 7.1, worldwide consumption of silicon wafers in 1988 was 1630 million square inches (MSI). This increased to 1705 MSI in 1991.

The onslaught of larger wafers will dramatically effect batch type reactors. The major batch reactor advantage of throughput will be lost. The hexode wafer loading of 24 four-inch wafers was reduced to 18 with five- and six-inch wafers. Loading will decrease to 10 eight-inch wafers and the system will be correctly termed a "pentode".

Single wafer systems are now capable of processing in excess of 40 four-inch and 20 eight-inch wafers per hour. As wafer sizes increase, the throughput advantage of the hexode reactor decreases. As batch and single wafer system throughputs approach equality, various other factors become major issues.

Table 7.1. Silicon wafer size usage in 1988 and 1991. (millions of square inches—MSI)

Wafer size (in inches)	MSI 1988	MSI 1991
3	39	21
4	632	425
5	452	558
6	500	691
8	7	10
TOTAL	1630	1705

PLASMA ETCHING

Today and tomorrow, VLSI device manufacturers will insist on etching equipment that addresses the following issues:

- Wafer transport automation
- Low particulate contamination
- Precise endpoint detection with multiple endpoint capability
- Microprocessor control with central computer remote monitoring and control
- Design simplicity
- Class 10 cleanroom compatibility
- 90% plus uptime to manufacturing
- Robotics interfacing
- Wafer sizes easily changed to prevent obsolescence

The hexode reactor, as well as various other batch etch systems designs, will have difficulty competing with single wafer systems if all the above requirements are to be met. Wafer transport automation is certainly not an impossibility for batch reactors. However, design simplicity becomes a major factor when competing against single wafer systems. Load locks, transport arms, and wafer pick-up devices all play an important role in design simplicity and feasibility.

A reduction in the levels of particulate contamination are dependent upon both mechanical and chemical considerations during the wafer transport and etching processes. Hardware design and particle filtration are factors that impact contaminant levels. Gas distribution and purge systems, air flow about the reactor, welding and tube cleaning procedures, and mechanical transport mechanisms must all be considered as important user issues.

Not only is it important to keep particulates and contaminants to a minimum within the etching system, but external controls are necessary to improve yield. The next generation of etching systems must be Class 10 cleanroom compatible and have mechanical robot interfacing. Operator-related contamination must be eliminated. Automation to and from the reaction chamber will increase yield, and possibly throughput, by the elimination of operator-related contamination.

Wafer transport designs play an important role in particulate levels that are machine contributable. Ideally, moving parts, meshing gears, transport chains, and motors must be kept out of the vacuum system. If residing in the load locks or reaction chamber, they must be kept to a minimum and below the plane of the wafer surface. Wafer edge pickup is not acceptable due to par-

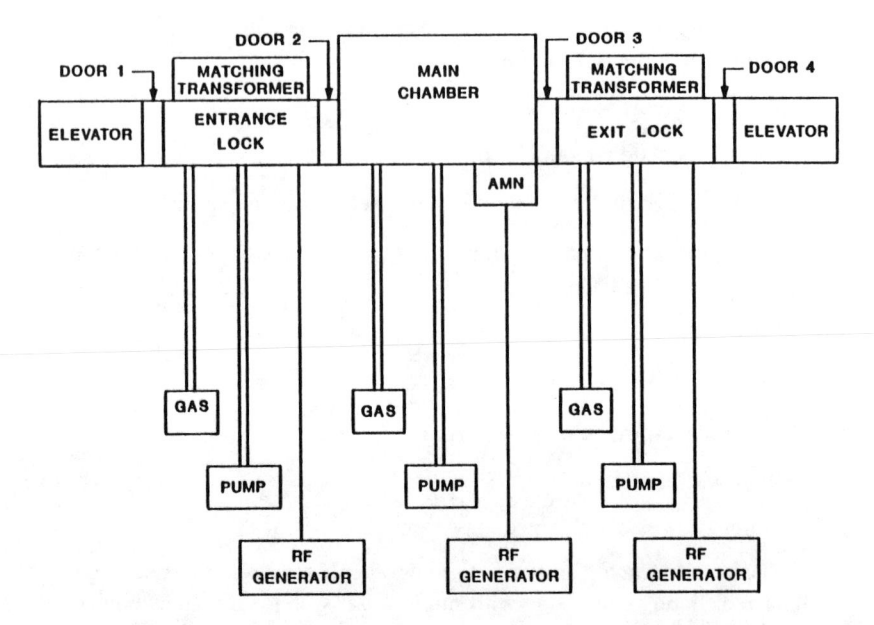

Figure 7.1. Schematic of typical single wafer RIE system.

ticulate formation from possible photoresist residues around the edge of the wafer. Edge bead removal equipment may help this situation.

Single wafer processing systems, shown schematically in Figure 7.1, lend themselves to complete cleanroom automation much more easily than batch type reactors. The semiconductor industry itself is moving toward the inline concept through factory automation, robot interfacing, and the elimination of operators in the cleanroom environment.

Single wafer endpoint detection has tremendous advantages over batch endpoint detection. In a batch system, endpoint detection is taken as an average of all the wafers in the reactor. This severely limits accuracy. Single wafer systems are able to endpoint each wafer individually, thus improving reproducibility and uniformity from wafer-to-wafer.

As devices become more compact and sophisticated, the etching process follows suit. Multi-layer materials are becoming commonplace. As a result, endpoint detection is becoming more and more critical. Multiple endpoint capability is necessary in order to etch stop on a designated layer. Batch type "average" endpoint detection is not capable of satisfying such a requirement. Once again, as wafer costs increase, single wafer endpoint detection becomes much more critical.

PLASMA ETCHING

Batch systems compound the problem of individual, accurate endpoint detection since wafer loading effects are inherent in all batch reactor designs. As wafer loads increase, throughputs decrease. Also, loading effects change etch rates. Both run-to-run and across-the-wafer etch rates change. This is why an end-point detector for batch reactors sees an average endpoint at best.

Single wafer systems offer yet another unique feature. In-line, multi-chamber configurations are easily designed and utilized. This concept gives rise to an important consideration for processes that require changes in chemistry. For example, the aluminum etching process requires a post etch treatment for the prevention of corrosion. The bulk aluminum etch step itself requires the use of chlorine chemistry, while the passivation step utilizes fluorine chemistry. Separate, distinct process chambers operating simultaneously, yet independently, are capable of achieving the desired process results. A bonus of increased throughput and absence of cross chemistry contamination are also attained. This concept is presently in its infancy but used by most single wafer etch system manufacturers in their aluminum etch system design. The possibilities are endless as more complex devices become more challenging to etch.

In the early 1980s, Drytek, acquired by General Signal, developed a multi-chambered etching system. This unit, known as the Quad, is comprised of four distinct process chambers. Each chamber is fully process independent of the others. However, each chamber shares one common transport mechanism. The transfer arm is a throughput bottleneck. Since all four chambers can process at the same time, theoretically all four chambers can finish processing at the same time. This slows throughput considerably since each wafer must wait its turn for transfer. However, the elimination of cross chemistry contamination and the promise of multi-layer processing remain desirable features. Electrotech followed suit with the four-chamber PLASMAFAB M4 and two-chamber Omega 2. Applied Materials' Precision 5000 Etch has two standard chambers that can be expanded to four as shown in Figure 7.2.

Advanced processing becomes much more complex, if not impossible, in batch type reactors. Transfer from process chamber to process chamber requires the use of vacuum load locks. This type of design on batch reactors increases the complexity of the system beyond desirability.

An additional design consideration that greatly affects total system uptime is reactor cleaning. Cleaning cycles can be divided into two sections. One being the time it physically takes to clean the reactor, the other being the time it takes to access the reactor. Another important factor is the extent to which cleaning is required. Many systems on the market today require cleaning of the transport mechanism, vacuum system, lock doors, and various peripherals in addition to the reactor itself. Obviously, this adds dramatically to downtime. Neither batch nor single wafer systems have eliminated the neces-

150				PLASMA ETCHING

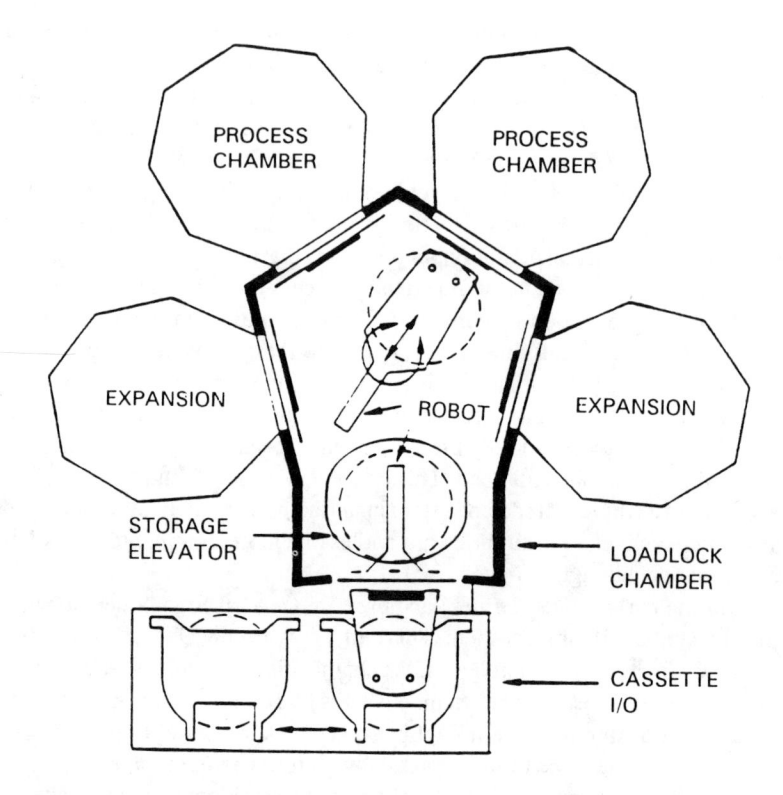

Figure 7.2. Schematic of precision 5000 Etch multichamber system.

sity for the cleaning cycle. However, dependent upon the manufacturer, the cleaning process can be minimized to as little as 30 to 45 minutes.

Atmospheric contaminants are a major source of process non-repeatability. Water vapor and oxygen are primary offenders. The use of vacuum load locks virtually eliminates these unwanted materials from entering the reaction chamber. Single wafer systems lend themselves to load-lock designs via the inline concept. Once again, batch systems require greater design complexity if outfitted with vacuum locks.

Some of the advantages of single wafer etching systems are:

- Single wafer endpoint detection leads to improved wafer-to-wafer uniformity since only one wafer is monitored

- Automated load/unload cycles reduce contamination and wafer breakage

- Elimination of multiple wafer loading effects during process

PLASMA ETCHING

- Load locks reduce atmospheric contaminants
- Improved run-to-run reproducibility
- Multiple chamber processing capability for advanced device manufacture
- Increased operator safety through the use of load locks
- Single wafer processing lends itself to automation and the concept of eliminating human intervention in the cleanroom
- Design simplicity is more easily achieved since the system is processing only one wafer at a time

Batch systems, most notably the hexode reactors, are capable of achieving very high throughputs in excess of 60 wafers per hour for certain etch processes. Throughput is the most cited advantage. Unfortunately, batch systems have the following drawbacks:

- Endpoint is determined by the average of all wafers in the reactor, thereby reducing accuracy
- Wafer loading effects change etch rates
- Manual loading/unloading increases particulate generation and the possibility of wafer breakage
- Exposing the reactor interior to atmosphere prior to each run increases water vapor concentrations and contamination
- Opening the reactor exposes the operator to potentially toxic substances
- Batch automation does not lend itself to design simplicity

Batch systems dominated the worldwide plasma etching equipment market through 1989. Reasons for market dominance included automation, high throughput, present design rules, and large, tolerant process windows.

Planar batch systems have been in existence for over a decade and are well understood. The hexode reactor grew from the knowledge of planar systems. Batch systems require relatively low power levels during process and are therefore less likely to produce substrate radiation damage. Present design rules do not yet demand the use of single wafer systems, even though they exhibit advantages over batch systems.

Batch system automation has improved dramatically over recent years. Today's systems generally exhibit clean designs that function well in production environments. Many batch systems now have vacuum load locks to eliminate reactor exposure to the atmosphere.

152 PLASMA ETCHING

Although their market share continues to decrease, batch processing systems will remain dominant in R&D and limited production capacities.

7.1.2. RIE Versus PE

Production of today's VLSI circuitry places specific requirements on plasma etching processes. Among these requirements are the following:

- Anisotropic etch profiles
- Anisotropic etch profiles with controlled taper capability
- Controlled endpoint detection
- Etch uniformities better than +/–5%
- Etch resolution better than 2 micrometers
- Selectivities greater than 4:1

Two of the major types of dry processing equipment are Plasma Etching and Reactive Ion Etching, as shown in Figure 7.3.

Plasma etching is routinely defined as a chemical inter-action between the reactive gas species in the plasma and the substrate. The substrates are placed on an electrically grounded electrode. Reactive species diffuse through the dark space above the wafer and etch the substrate. Volatile by-products are pumped away through the vacuum system. Plasma etching is characterized by high etch rates and high selectivities. However, high operating pressures (1 Torr) result in a low ion mean free path and resist undercutting occurs. As a result, plasma etching is also characterized by its omnidirectional etch profiles.

On the other hand, reactive ion etching is considered to exhibit both chemical and physical properties. The substrates are placed on the RF powered electrode. Operating pressures are lower (0.1 Torr) than in the PE mode so that a greater potential drop between the plasma and electrode results in a high degree of etch directionality. The increased momentum of the reactive gas species produces a physical sputtering of the substrate material. This results in very little undercutting of the resist profile. As a result, reactive ion etching is also characterized by its anisotropic (straight wall) etch profiles.

Typical characteristics of PE and RIE are illustrated in Table 7.2.

To be successful, both etching modes require proper selection and careful control of operating parameters, such as choice of gases, pressure, flow, power, temperature, and electrode spacing.

Etching of a thin film or substrate material occurs through the chemisorption of reactive gas species and the formation of a volatile compound.

PLASMA ETCHING

The energy of the incident species on the surface of the substrate initiates the reaction. Physical etching components, as in RIE, involve the transfer of momentum from the bombarding species to the atoms of the substrate surface. As the momentum vector is directed away from the surface, the ejection of a surface atom results.

Radiation and high energy bombardment are necessary in dry processing; especially in reactive ion etching. Energetic ions and high energy photons can produce electron traps and neutral centers in the film during processing. Trapped holes can be removed by annealing at higher temperatures. Neutral centers can also be annealed at much higher temperatures if the device can withstand it. Radiation damage is not as severe in the PE mode since the wafers are mounted on electrically grounded surfaces.

Ion acceleration in RIE is the result of developed DC voltages on the powered electrode. Typical developed DC voltages range in the hundreds-of-volts region. The developed DC bias accelerates the ions from the plasma to the substrate. This ion bombardment in the RIE mode contributes greatly to the concern of device radiation damage (Oehrlein *et al.* 1986).

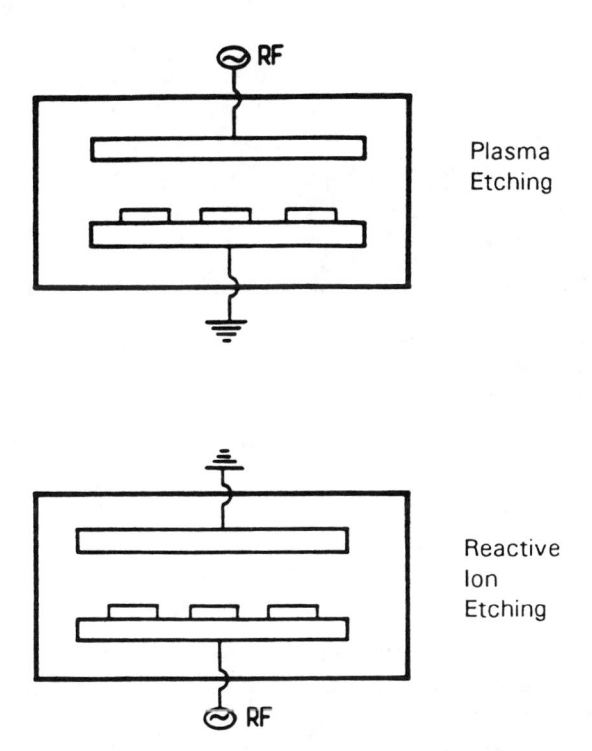

Figure 7.3. **Schematic of dry processing systems.**

Table 7.2. Characteristics of dry processing systems.

	Plasma etching	Reactive ion etching
Pressure (Torr)	1–0.1	0.1–0.001
Flow (SCCM)	100	10
Mechanism	Chemical	Chemical/physical
Active species	Atoms, ions, radicals	Ions, radicals
Selectivity	Good	Reasonable
Edge profile	Isotropic/anisotropic	Anisotropic
Location of substrate	In plasma on ground electrode	In plasma on powered electrode
Profile	Isotropic etching	Anisotropic etching
Rate	High etch rate	Low etch rate
Selectivity	Good selectivity	Reasonable selectivity
Throughput	High throughput	High throughput
Detection	End-point detection	End-point detection
Radiation damage	Little radiation damage	Radiation damage
Features	Etch residue	Etch all materials

Because of its characteristic anisotropic etch profile, reactive ion etching is considered necessary for etching present and future device geometries. Plasma etching is not capable of etching the fine lines required for today's technology. The inherent disadvantages of PE processing (i.e. omnidirectional etching) restrict it to larger device dimensions.

7.1.3. Impact of Electron Cyclotron Resonance

Electron cyclotron resonance (ECR) is a condition whereby a resonant field with a strength of 875 Gauss is used in the generation of a plasma from a microwave source at 2.45 GHz.

Four developments in ECR technology have progressed concurrently:

- In Japan, NTT Technology has patented a system illustrated in Figure 7.4. The technology has been licensed to NEC subsidiary Anelva, Mitsubishi, Hitachi, Tokyo Ohka, Sumitomo, and MRC. Lam has a license from Sumitomo to market the system in North America and Europe. The system uses biased grids to parallelize the beam.

- The Centre National D'Etudes des Telecommunications (CNET) and Centre National de la Recherche Scientific (CNRS) have co-developed a system with Electrotech. This system uses a multipolar approach incorporating a single-source, high-strength magnetic field, as shown in Figure 7.5.

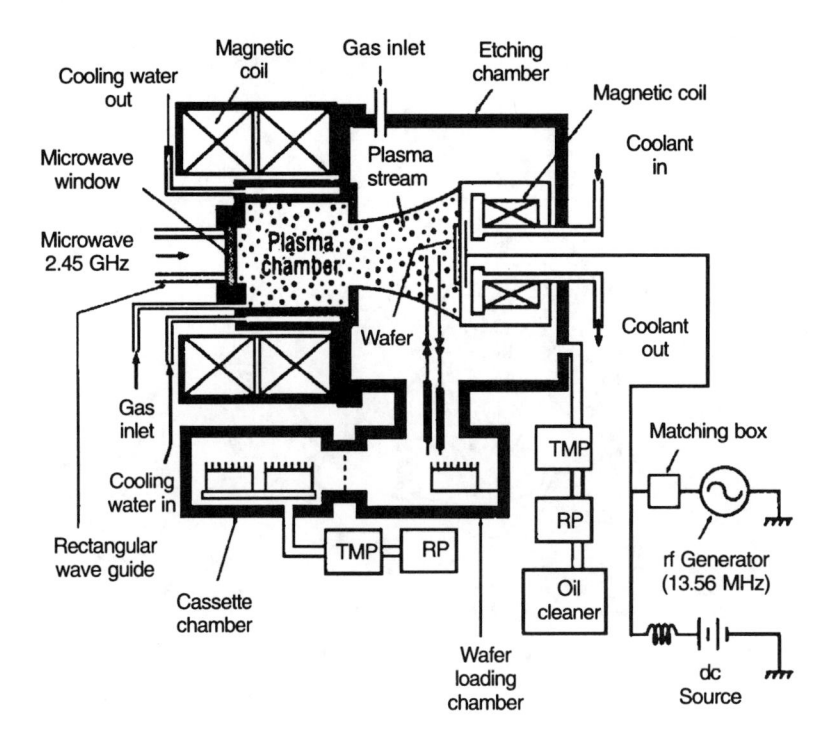

Figure 7.4. Schematic of beam-source ECR.

- The Alvey ECR Project, a joint project funded by the UK government with Oxford Plasma Technology. This method of confinement involves the interaction of the microwaves and permanent magnets with an rf bias completely separate to the plasma generation.

- Leybold AG's ECR source is part of a compact, modular reactor that can handle 200 mm wafers and is suitable for use in future multichamber processing systems.

The frequency of ECR systems at 2.45 GHz, compared to 13.56 MHz of conventional systems, improves the cracking efficiency by a factor of three, leading to more ions per cubic cm with a greatly reduced energy: this results in higher selectivity.

Shown in Table 7.3 is a comparison of RIE, beam-source ECR, and multipolar ECR.

Beam-source ECR, also referred to as "classical" ECR, uses a controlled magnetic field strength, enabling electrons to orbit magnetic flux lines in step

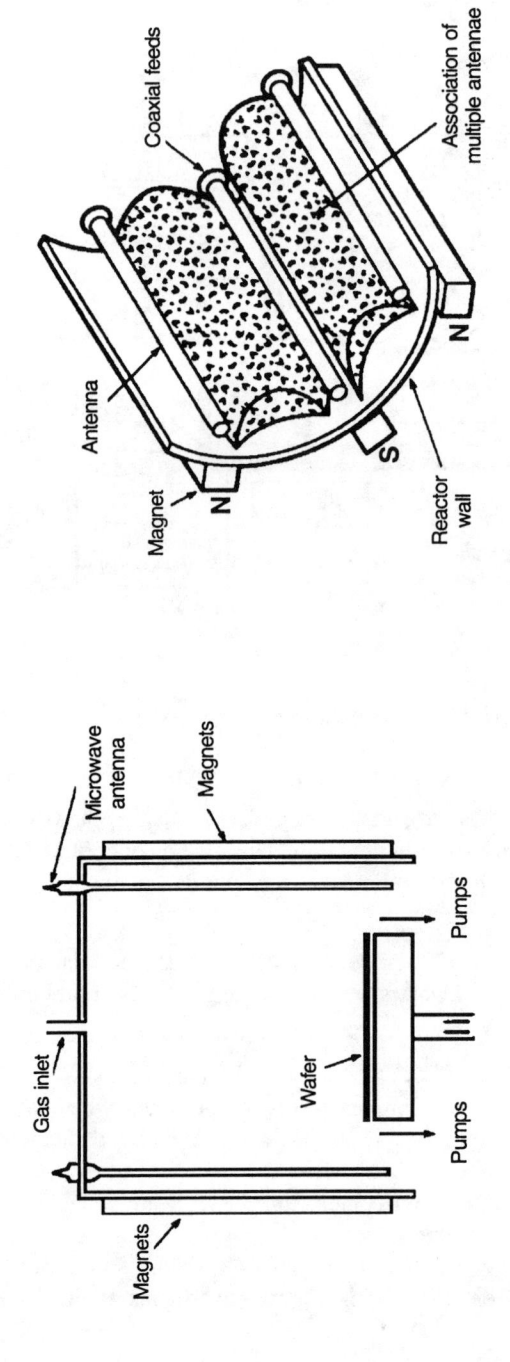

Figure 7.5. Schematic of multipolar ECR.

Table 7.3. **Comparison of RIE and ECR.**

	RIE	Beam-source ECR	Multipolar ECR
Plasma type	rf diode	remote plasma	triode
Pressure (mtorr)	5–100	0.1–1	0.1–5
Ion current (mA/cm^2)	0.1–1	0.5–1	1–3
Ion energy (eV)	400 - 1000	40 - 1000	40 - 200
Grid-free operation	Yes	No	Yes
Magnetic enhancement	No	Yes	Yes
Magnetic source	–	high current coils	permanent magnets
Field line direction	–	intersects wafer	parallel to wafer
Current independent of energy	No	No	Yes
Uniformity	Good	Fair	Excellent

with the 2.45 GHz microwave excitation. The result is an intense, high-density plasma that can be maintained a low pressures of 10^{-5} to 10^{-2} (Asmussen 1989).

A disadvantage of this approach is that the plasma stream is divergent, causing non-uniformities in plasma density and ion direction. The density of the plasma often decays as it flows beyond the source to the region where the wafers are etched. Scattering can occur when source and wafer are separated by a long diffusion zone of falling density (Cook and Donohoe 1991).

With this design, wafers are usually maintained downstream of the plasma to limit exposure to the intense discharge. They can also be biased to control the ion bombardment.

The multipolar ECR, also referred to as "distributed" ECR (DECR), operates in a similar fashion to the low-pressure, magnetic dc multipole reactor, whereby an intense plasma exists close to the walls of the vertical cylinder and a single wafer is maintained centrally and toward the bottom of the cylinder (Limpaecher and MacKenzie 1973).

There are no major U.S. manufacturers of ECR systems, although Lam is marketing the system in the U.S. produced by Sumitomo in Japan, and Oxford Plasma Technology markets its system from its parent company in the UK.

Although introduced in R&D labs in 1987 and in Japanese fab lines in 1988, ECR sales increased dramatically in 1989. However, Hitachi is the one of the only companies with an ECR used in production. Japanese companies are purchasing more ECR systems for use on 4 and 16 Mbit production lines.

Although ECR promises to be an evolutionary tool for the delineation of fine-line geometries, there have been some problems in optimizing process parameters. Anelva, for example, suspended marketing of its DECR system in mid-1988 because of technical problems—specifically because of sidewall etch and deposition. Delays in development of side-wall rf bias technology, a solution to the tendency of ECR technology to build uneven side walls, has been the major factor. Another problem has been delays in the development of a reliable automatch and autotune circuitry to link the power supply to the cathode.

7.1.4. Other Sources

After initial introduction of plasma etching systems in the 1970s, equipment manufacturers began introducing different types of reactor designs in the mid-1980s. Examples of these advanced systems are shown in Figure 7.6 (Flamm 1991), and include triode designs from Tegal and GCA, and magnetically enhanced parallel plate processors initially from Anelva, MRC, and Tokuda.

These systems were designed to control plasma generation and ion bombardment energy by separating them using two specially separate regions. Magnetically enhanced reactive ion etch (MERIE), is a low-pressure RIE method that runs at 10^{-3} Torr (Lin *et al.* 1984). At this pressure, the ion density is low, significantly affecting the etch rate. Therefore magnets are used to increase the ion density by controlling the geometry of the electrons producing the ions.

Tylan/Tokuda uses permanent magnets to achieve the effect in their HiRRIE 500 system. The magnets, on top a standard RIE chamber, are moved back and forth to create the dense plasma. This procedure increases Cl radicals by a factor of 20 over conventional methods.

Applied Materials replaced its hexode reactor with the Precision 5000, which uses software-controlled electromagnets to produce a magnetic field. The electromagnetic field provides the ability to vary the magnetic flux as well as rotate the field and vary the rate of rotation to optimize the process.

The helicon whistler source (Boswell and Henry 1985) (Perry, 1989), shown in Figure 7.7, and the helical resonator source (Cook *et al.* 1990), shown in Figure 7.8 are two new high-density rf plasma etching methods.

The helicon whistler operates by initiating whistler modes, which are stable in specific ranges of magnetic field and power. The helical resonator source employs a slow wave structure to achieve quarter or half wave resonance in a small volume. No magnetic field is required, as in the helicon whistler source.

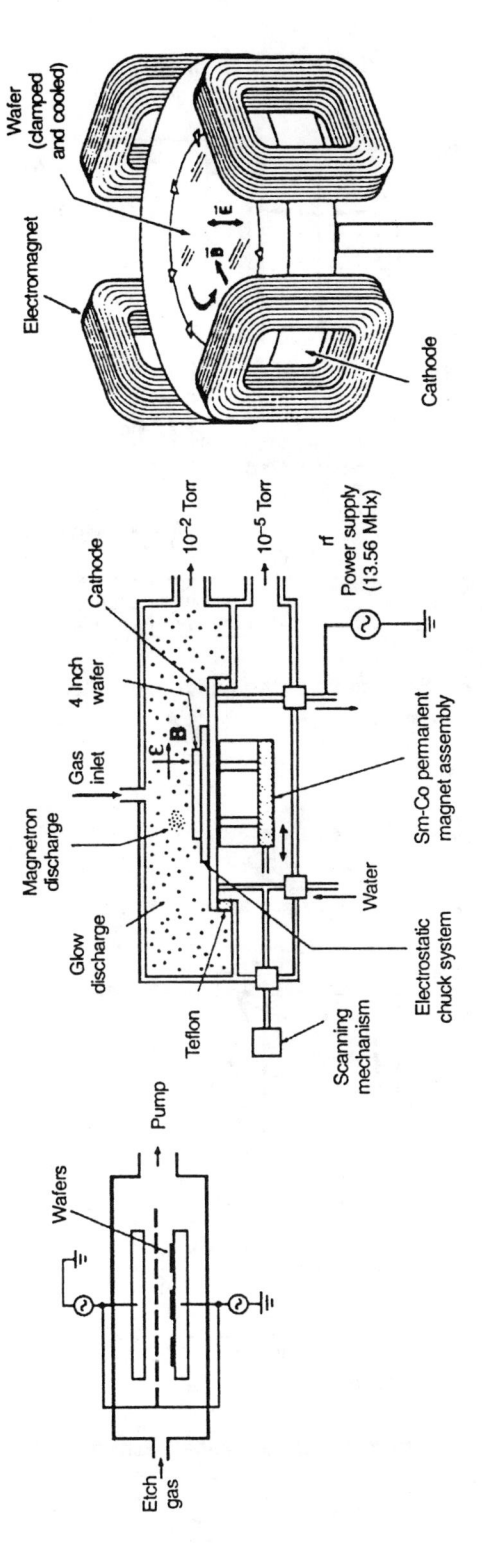

Figure 7.6. Various enhanced designs (a) triode, (b) HIRRIE, (c) MERIE.

160   PLASMA ETCHING

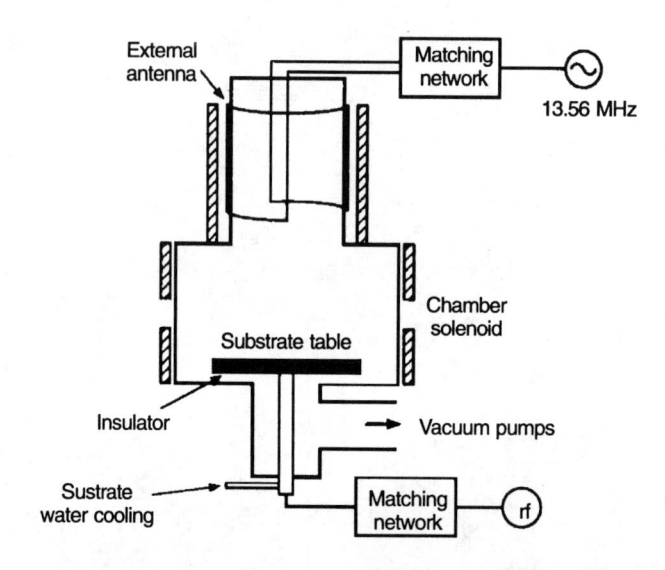

Figure 7.7. Schematic of the helicon whistler source.

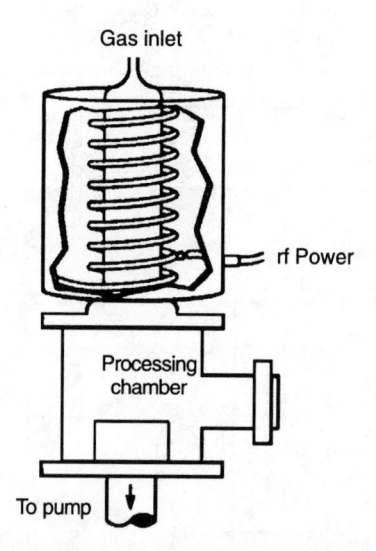

Figure 7.8. Schematic of the helical resonator source.

PLASMA ETCHING

Advantages of both sources include:

- Operation at 13.56 MHz, eliminating microwave health concerns
- Do not require large magnets
- Do not require microwave waveguides with specialized matching circuits
- Do not require complex field coils

Further designs are on the horizon, and their utility for submicron resolution remains to be proven. The enhanced systems discussed above are adequate at 0.75 µm, and could be extended further. The production of 256 Mbit devices in 1997 will require not only an understanding of the limitations of current and future designs, but an understanding of the plasma chemistry as well as the plasma physics of the reactor. Plasma chemistries of different processes are described below.

7.2. PROCESSING ISSUES

Plasma etching and reactive ion etching have been the subject of several review articles (Flamm and Donnelly 1981), (Muray and Brodie 1982), and (Coburn 1982). These have tended to describe the physical and chemical aspects of dry processing techniques in relationship to etchant and apparatus selection.

7.2.1. Chlorine Versus Fluorine Processes

As a general rule, chlorine based gases are considered much more difficult to handle due to their corrosive nature. Once introduced into the plasma system, they can have a devastating effect on chamber seals, moving parts and construction materials.

Dry processing systems are generally more tolerant of fluorine chemistries and process-gas-associated hardware problems are minimal. However, gas handling is an important issue to consider for both chlorine and fluorine based process gases.

Choice of design materials may also be effected by process gases. Reactor materials, O-rings, transport mechanisms, viewport designs, vacuum systems, and gas-handling schemes are all to be considered in view of the process chemistries to be used.

Aluminum, aluminum alloys, polysilicon, single crystal silicon, silicides, silicon dioxide, silicon nitride and polyimide are among the common materials that etch readily in a dry etch reactor. Each film has its own characteristic etch qualities. Choice of etchant chemistry and reactor type (RIE or PE) both influence the dynamics of the plasma interaction with the film surface. This interplay of processing parameters and system characteristics has a pronounced effect on the resultant etch characteristics.

Aluminum etching requires a two step process, the removal of the native aluminum oxide film, and the bulk metal etching. Selection of the proper gases is important since aluminum oxide does not etch using the same gas as the bulk aluminum. Chlorine readily removes aluminum but is ineffective in removing the oxide layer (Schaible *et al.* 1978). BCl_3, $SiCl_4$, and CCl_4 are all effective in removing the native aluminum oxide layer. BCl_3 and $SiCl_4$ also provide an additional bonus; they are both an excellent getter of water vapor and oxygen (Tokunaga and Hess 1981). CCl_4 is not a good getter of water vapor or oxygen and tends to form excessive polymer buildup within the reaction chamber. Additionally, very few semiconductor manufacturers allow the use of carbon tetrachloride in the fab since it is a suspected carcinogen.

After native oxide removal, the bulk aluminum etching is achieved using chlorinated gases. Blending of gases to alter their individual etch characteristics is a common technique. The most commonly used chlorinated gases are BCl_3, Cl_2, or $SiCl_4$. Inert constituents such as argon, nitrogen, or helium are added as dilutents with specific properties. Chlorine will etch aluminum without the assistance of ion bombardment. This suggests that a purely chemical mechanism is responsible. Therefore, the inerts are needed to slow down the reaction and make it controllable.

Reactive ion etching is the method of choice for metal etching. Since chlorine will etch omnidirectionally without ion assistance, undercutting is a common result. RIE has the capability to direct the ions to achieve an anisotropic profile. Sidewall protection also contributes to anisotropy. Any polymer formed by C–H bonds will serve to form a passive inhibiting layer that is easily removed from horizontal surfaces. Photoresist degradation during processing also contributes to sidewall protection.

Fluorocarbons and fluorine compounds should not be used for aluminum etching since they form non-volatile fluorides.

Typically, aluminum films are alloyed with 1% to 2% silicon to prevent junction spiking and 2% to 4% copper to prevent electromigration. The silicon component forms $SiCl_4$ in chlorine plasmas, which is volatile and readily removed during processing. Copper, however, does not form volatile chlorides at processing conditions (Schaible and Schwartz 1979). Therefore, reactive ion etching is used to sputter any copper residues during overetch conditions.

PLASMA ETCHING

Polysilicon may be etched using either chlorine or fluorine chemistries. Common fluorine etchants are CF_4 (Bondur 1979), SF_6 (Eisele 1981), and NF_3 (Eisele 1980). Etch profile is generally isotropic in the plasma etch mode due to the long lived fluorine species. Anisotropic profiles are obtained using reactive ion etching. The anode-to-cathode ratio in the reactor design plays an important part in controlling anisotropy. The larger the anode-to-cathode ratio, the more anisotropic the profile. Small anode-to-cathode ratios could cause some undercutting in the RIE mode.

Chlorine chemistry is also capable of etching polysilicon since the effluents are volatile. Chlorine is a fairly good etchant. However, it cannot etch through the native silicon oxide readily. High rate etching may be achieved if the substrates are pretreated with an HF, C_2F_6, SF_6 or NF_3 plasma (Mogab and Levenstein 1980).

The use of chlorine-containing halocarbons is preferable to pure chlorine because they are non-corrosive, not hazardous, and may produce process-tailored results, dependent upon the choice of gases (Flamm 1981). Chlorine gases generate short-lived radicals and ions, whereas fluorine's long lived species can migrate across the surface, etch in a lateral direction and result in undercutting (Winkler 1983).

Commonly, polysilicon is doped with phosphorous in order to reduce resistivity. Etch rate differences between doped and undoped poly are related to gas compositions. For instance, the addition of chlorine will increase the etch rate of doped polysilicon.

Silicon trench etching is a popular way isolating densely-packed devices and fabricating vertically-oriented capacitors and three-dimensional structures. As VLSI packing densities increase, individual devices must be isolated. Typically, a deep trench is etched in the silicon substrate then filled with a dielectric material. A highly anisotropic etch profile is desired, since it is important that very little wafer real estate be lost. This structure is illustrated in Figure 7.9. Applied Materials' Precision 5000 Etch has demonstrated a depth-to-width aspect ratio of 21:1, with the trench 0.2-µm wide and 4.2-µm deep.

Chlorine chemistries and the reactive ion etch mode are the most effective ways of meeting the depth and width requirements in trench etching. The plasma etch mode may also produce an anisotropic profile. The addition of a sidewall protectant during high pressure plasma etching helps achieve a straight wall profile at the cost of reactor contamination.

Once again, in the RIE mode, a high anode-to-cathode ratio is necessary to produce the self bias required for ion acceleration. Etching is accomplished with a mixture of gases: Cl_2 for etching; BCl_3 for scavenging silicon dioxide and oxygen (to eliminate artifacts of micromasking); and $SiCl_4$ to provide sidewall passivation (to prevent bowing and trenching caused by scattered ions).

Refractory metal silicides (titanium, tantalum, tungsten, and molybdenum) have a lower resistivity than doped polysilicon. This makes them more suitable as gate and interconnect materials. The silicides are typically used as a sandwich structure with polysilicon. The resultant structure, with the silicide deposited over the polysilicon, is termed a polycide. Polysilicon is used to prevent the refractory metal silicide from damaging the underlying thin gate oxide layer.

Dry processing parameters must be optimized to etch the sandwich layers anisotropically while maintaining good selectivities to the underlying oxide. The silicides and polysilicon possess vastly different etch characteristics. Severe undercutting results if the etch process is not properly tailored to the particular refractory metal silicide and the dopant level of the underlying polysilicon.

Ta, Ti, and Mo silicides may be etched in either chlorine or fluorine plasmas. However, tungsten silicide (WSi$_2$) can only be etched with fluorine gases since the chlorides of tungsten are non-volatile. It is important when using fluorine chemistries to stop at the silicide/polysilicon interface since fluorine will typically undercut polysilicon. CCl$_4$ will etch both layers anisotropically, but leaves a residue on the underlying silicon dioxide surface.

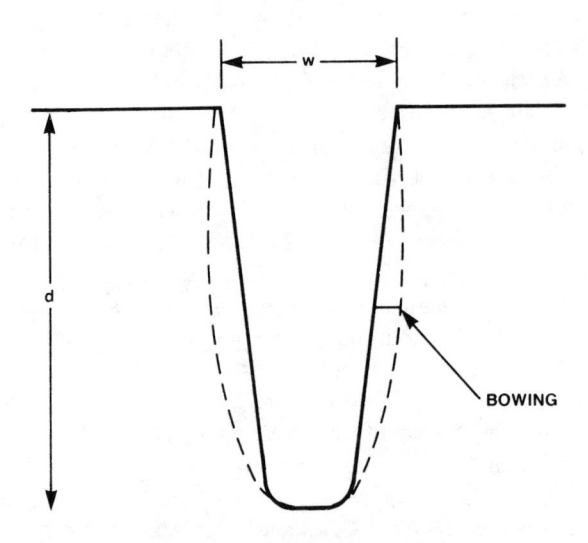

Figure 7.9. Silicon trench structure.

From this it can be determined that the dry etching parameters such as gas composition, power, pressure, flow, and mode of operation must be optimized to achieve vertical etch profiles in polycide structures.

Both SiO_2 and SiN are commonly used dielectrics and passivation layers in VLSI device manufacture. Anisotropic profiles for contact vias are not suitable. This is because vacuum deposited metals do not provide adequate coverage over vertical steps. A tapered sidewall vastly improves subsequent step coverage. Further, when etching to expose underlying silicon and polysilicon, it is important to maintain a high selectivity to Si.

More important than a sloped profile is bottom rounding of the via. A rounded profile at the vertical/horizontal intersection reduces the possibility of microvoids and cracks in the overlying film.

Silicon dioxide is etched by reactive species such as CF_2 and CF_3 (Heinecke 1975). Fluorocarbon plasmas can readily generate these species. However, if atomic fluorine is liberated, the selectivity of SiO_2 is greatly reduced since atomic fluorine is a good silicon etchant. The addition of hydrogen (Ephrath 1979) will remove fluorine as HF, a poor etchant of silicon. Hydrogen is most conveniently added as CHF_3 (Lehmann and Widmer 1978).

The addition of hydrogen has a tendency to deposit polymeric materials on the wafer surface, thereby adversely affecting etch rate. This phenomena is quite common at pressures greater than 250 mTorr. By operating at lower pressures (<100 mTorr) and utilizing ion bombardment, polymer contamination can be entirely eliminated.

Sloped wall oxide via etching can be achieved through controlled photoresist erosion techniques. The etch rate of the resist to the oxide can be adjusted to produce sloped profiles. With this method, the slope of the resist is effectively transferred to the oxide. It is possible to produce wall slopes in the range of 60 to 90°. This can be accomplished by adding O_2 or SF_6 (Bondur and Clark 1980).

Silicon nitride can be etched with fluorine ions and, due to fluorine's long lived activation, etches isotropically. Typical etch gases for silicon nitride are CHF_3 and O_2 or CF_4 and O_2. Process gas selection is dependent upon the underlying material. CF_4/O_2 mixtures have a greater selectivity to silicon dioxide, while CHF_3/O_2 combinations are used over silicon (Castellano 1984).

Polyimide is a dielectric material that is useful in multi-level interconnection technology. This thick material (1 to 2 µm) is spun on a wafer in a similar manner as photoresist. As it spreads over the wafer, it tends to planarize the surface by filling in irregular topography. The same equipment is used to spin-on both resist and polyimide. The resultant planarized surface is important for patterning additional layers, since focusing of lithographic equipment is more precise on a flat surface. The polyimide is then cured at about 400° C.

Via hole dry etching of polyimide is quite simple. The singular process gas, O_2 is used. Sloped walls can be achieved by sloping the photoresist mask.

7.2.2. Multilevel Structures

Integrated circuit complexities are increasing while device geometries are shrinking. VLSI design engineers are utilizing multilevel technology to increase device functions while keeping device "real estate" to a minimum. New materials are used in IC fabrication. Devices are changing their "skyline" by building up as multilayer structures.

With this increased complexity comes additional problems for process engineers and equipment suppliers alike. Linewidths are shrinking well below 1 μm, which points in the direction of advanced dry processing technology. Many processes must be improved in order to meet the challenge of these design techniques.

Wall profiles must be tightly controlled. Achieving anisotropic profiles with controlled tapers during dry etching can prove challenging to any process engineer. The problem is compounded when design rules stipulate little or no dimensional line loss during the etch process. Anisotropic profiles are easier to achieve with reactive ion etching, but ion damage is an important consideration. Excessive radiation can cause malfunctions in all or part of the device.

As the thickness of device layers decrease, selectivities to underlying materials becomes critical. Process parameters such as pressure, power, DC bias, gas composition, and flow must be precisely controlled. Mechanical and computer response times must be minimal.

To be certified for the production line, single wafer processing requires high etch rates without sacrificing throughput. For dry processing, that usually means a compromise in some other category. Typically, high throughputs are at the expense of high selectivities. However, automation, particulate, safety, and endpoint gains achieved by single wafer systems over batch systems are generally considered worth the compromise.

As previously discussed, single wafer systems are capable of multi-chamber processing and are limited only by design. Advanced multilayer processing often requires changes in basic chemistries. By using multi-chambered, single wafer systems for each process chemistry, it is possible to actually increase throughput over hexode systems, even for four, five, and six inch wafers. Also, each chamber may perform the same process, thereby increasing the throughput accordingly.

7.2.3. New Metallization Materials

Metal interconnections pose various problems for plasma process engineers. New techniques and materials are constantly explored to make the dry etch process more repeatable and easier. Many times, circuit design will require material changes or enhancements for proper operation.

Metal deposition becomes an easier task at higher temperatures. Elevated temperatures produce quality films with good step coverage. However, the higher the temperature during aluminum sputtering, the more difficult to etch the film.

As discussed above, aluminum is commonly alloyed with either silicon, copper, or both. The addition of these materials serves specific purposes. One to two percent silicon is added to prevent junction spiking, while two to four percent copper prevents electromigration.

Aluminum/silicon alloys will leave residues if not properly etched. These silicon residues are generally considered cosmetic in nature. However, if device problems arise, the residues may be removed by further processing.

Aluminum/copper alloys are much more difficult to etch. Copper does not form volatile chlorides at dry etching conditions. Therefore, copper residues remain after processing and can cause poor device performance. The use of reactive ion etching to sputter the copper is quite successful. However, the higher the concentration of copper, the more difficult to reactive sputter etch. Titanium is currently being utilized as a replacement for some of the copper. This material does have volatile chlorides.

The use of an underlying titanium/tungsten (TiW) barrier layer provides process challenges. The TiW layer is typically <1,000 angstroms thick. TiW etches easily in fluorine chemistries using CF_4. Aluminum etching is accomplished with chlorine gases. However, post-etch corrosion is prevented by exposure to fluorinated plasmas. Already, process problems are apparent. The underlying TiW layer can be severely undercut during the fluorinated passivation step. Careful selection of process parameters and etch chemistries enable the process engineer to effectively passivate the wafer surface and prevent TiW undercutting.

7.2.4. GaAs Processing

Gallium arsenide substrates are not easily etched in single wafer or hexode etching systems. The problems encountered are related to wafer handling. Once the wafer is in position on the electrode, it may be readily etched in chlorine gases. Gallium arsenide wafers are extremely fragile and expensive. Most users are still processing two- and three-inch wafers with small, parallel plate, batch type reactors.

Gallium arsenide is not readily etched in fluorine-based chemistries due to the formation of involatile GaF_3. However, it is easily etched in chlorine gases. By blending gas mixtures, a number of processes that exhibit a wide range of etch characteristics may be developed. Typical process gases include CCl_2F_2, Cl_2, and $SiCl_4$. The addition of inert constituents such as Ar or N_2 helps to control the reaction since etching in chlorine gases proceeds so rapidly (Donnelly et al. 1982).

Utilizing the RIE mode of operation takes advantage of ion bombardment that allows anisotropic etching (Cooper et al. 1987). This technique is particularly useful for through-the-wafer via etching. Process requirements are quite stringent. The typical etch depth needed is >100 μm. Profile must be anisotropic with a controlled positive taper. The crystalline structure of GaAs severely restricts achieving the required profile without the use of a sidewall protectant. A thick photoresist mask will degrade sufficiently during the process to provide adequate masking of the vertical etched surfaces without forming excessive polymer. If oxide or nickel masks are provided, the addition of a polymer forming gas will provide the necessary sidewall protection during etch.

Through-the-wafer gallium arsenide via etching is not particularly suited to single wafer processing due to long process times. Via etch rates are dependent upon the diameter of the via and the etch depth requirement. As the depth approaches 50 μm, etching slows considerably. A typical through-the-wafer via of 2 mils diameter and 100 μm etch depth requires approximately 2 to 3 hours processing time. Obviously, much greater throughputs may be achieved by using batch systems. The parallel plate design batch systems are best suited to GaAs processing since wafers need not be clamped or held down causing concern for substrate breakage.

7.3. SAFETY ISSUES

7.3.1. System Design Considerations

User safety is of prime importance. Vendors must realize that personnel safety must be designed into the equipment. All suppliers must take it upon themselves to provide the safest hardware and processes available. Users must install all plasma processing equipment according to the vendor's specifications. Of course, vendors must supply the proper documentation for proper installation and system maintenance.

Many safety considerations are design oriented. Once received, it is extremely difficult to modify a system to account for design shortcomings. The following is a general list of universal safety considerations that should be provided on all plasma processing equipment.

PLASMA ETCHING 169

- Mainframe exhaust
- Separate gas box exhaust
- Chamber exhaust for reactor cleaning
- Follow national electrical codes
- Provide safety shields over moving mechanical parts
- Provide electrical and mechanical interlocks where needed
- Insulate hot and cold exposed surfaces
- Provide easy access to all subassemblies
- Furnish detailed service documentation and drawings with every machine
- Make provisions for quick, easy, access to parts that require regular maintenance or replacement
- Do not use cryo traps or pumps for process gases

Additionally, suppliers can enhance process safety by providing plasma processes that require minimum use of extremely hazardous gases. For example, substituting another chlorine-based gas for CCl_4, which is a carcinogen.

7.3.2. Gas Handling

Many etch gases are either corrosive, toxic, or both. If not toxic themselves, many contain some level of toxicity from trace gases during manufacturing or packaging. For example, phosgene is present in boron trichloride. Carbon tetrachloride is a suspected carcinogen, while chloroform will surely achieve the same carcinogenic status. Silicon tetrachloride and silicon tetrafluoride hydrolyze to form hydrochloric and hydrofluoric acids respectively.

Volatile etch by-products have been found to contain irritants and toxins. Gases such as carbon tetrachloride form polymer chains under plasma conditions. These polymers form various substances that are capable of causing physical damage to the nervous system, internal organs and various epidermal layers.

Personnel safety is greatly enhanced by vacuum load locks. The locks act as buffer zones and isolate the hazardous gases in the main reactor so that operator exposure is not an issue.

Safety from exposure to source gases is of critical importance. The process gas handling system has great potential for disaster. Many of the gases used in dry plasma processing are undetectable in sight and smell. They are silent killers.

The following are general gas handling considerations:

- Exhaust the system gas box
- Install alarms within the gas box, cabinets, and along tubing routes
- All tubing runs should be stainless steel with welded VCR fittings
- Leak check all lines to 1×10^{-9} sccm He
- Make use of only the finest stainless steel regulators
- Provide bellows sealed shut-off valves where needed
- Install an inert gas purge line to the bottle regulator
- Isolate gas line components (mass flow meters) with sealed shut-off valves upstream and down-stream for future replacement
- Design all gas lines with the ability to evacuate the line to the bottle
- Place all bottles in vented cabinets
- Do not place non-compatible gases in the same cabinet
- Vent reactive gases to a scrubber

All personnel must become intimately familiar with the properties of the gases they are handling. There is no excuse for ignorance of gas handling procedures. Perform maintenance and inspections regularly. Do not take chances for yourself and those around you.

7.3.3. Reactor Cleaning

Various polymers are formed and deposited within the plasma reactor during processing. These polymers must be removed on a regular basis to prevent excessive particulate buildup. The choice of process gas plays an important role in polymer buildup and reactor cleaning. Process parameters should be optimized so that reactor cleaning is a quick, easy task. For instance, metal etching may be performed with either $SiCl_4$ or BCl_3 and chlorine.

The $SiCl_4$ process leaves behind a tough polymer that must be scrubbed with scotch brite. Boron trichloride chemistries form a water soluble polymer that may be removed with a damp cloth. The BCl_3 process may be performed quicker, hence exposing the operator to noxious fumes for shorter periods of time.

PLASMA ETCHING

Reactor cleaning should only be performed while wearing proper safety attire. Gloves, safety glasses, protective apron, and mask should be worn while in the vicinity of the reactor during cleaning.

The area over the chamber should be exhausted. It is important to keep the air moving and exhausted away from personnel.

If possible, remove parts from the reactor to clean under a fume hood. Swap out the parts with clean ones. This hastens the cleaning process and minimizes exposure time. Dispose of all cleaning materials properly.

While evaluating equipment, users should pay close attention to reactor cleaning times and procedures. It is important to the safety of personnel that the procedure be quick and easy.

REFERENCES

Asmussen, J., 1989: *J. Vac. Sci. Technol.* **A7**: 883.
Bondur, J.A., 1979: "CF$_4$ etching in a diode system," *J. Electrochem. Soc.* **126** (2); 226–231.
Bondur, J.A. and H.A. Clark, 1980: "Plasma etching of SiO$_2$ profile control," *Solid State Technology* **23** (4): 122–128.
Boswell, R.W. and D. Henry, 1985: *Appl. Phys. Lett.* **47**: 1095.
Castellano, R.N., 1984: "Profile control in plasma etching of SiO$_2$," *Solid State Technol.* **27** (4): 203–206.
Coburn, J.W., 1982: "Plasma assisted etching," Proc. Tutorial Symposium on Semiconductor Technology, The Electrochemical Soc., Vol. 8255, pp. 177–217.
Cook, J.M., D.E. Ibbotson, and D.L. Flamm, 1990: *J. Vac. Sci. Technol.* **B8**: 1.
Cook, J.M. and K.G. Donohoe, 1991: "Etching issues at 0.35 μm and below," *Solid State Technology* **34** (4): 119–124.
Cooper, C.B., M.E. Day, C. Yuen, and S. Salimian, 1987: "Reactive ion etching of through-the-wafer via connections for contacts in GaAs FETs," *J. Electrochem. Soc.* **134**: 2533.
Donnelly, V.M., D.L. Flamm, and D.E. Ibbotson, 1982: "Plasma etching of III-V compound semiconductors," *J. Vac. Sci. Technol.* **A1**: 626.
Eisele, K.M., 1980: "Plasma etching of silicon with nitrogen trifluoride," Spring Meeting Electrochemical Soc., St. Louis, MO.
Eisele, K.M., 1981: "SF$_6$ a preferable etchant for plasma etching silicon," *J. Electrochem. Soc.* **128** (1): 123–126.
Ephrath, L.M., 1979: "Selective etching of silicon dioxide using reactive ion etching," *J. Electrochem. Soc.* **126** (8): 1419–1421.
Flamm, D.L., and V.M. Donnelly, 1981: "The design of plasma etchants, Plasma Chem. and Plasma Processes," *J. Vac. Sci. Technol. B* **1** (1): 23–30.
Flamm, D.L., 1991: "Trends in plasma sources and etching," *Solid State Technology* **34** (3): 47–50.
Heinecke, R.A.H., 1975: "Control of relative etch rates of SiO$_2$ and Si in plasma etching," *Solid-State Electron.* **18**: 1146–1147.

Lehmann, H.W. and R. Widmer, 1978: "Composition variations as a function of ejection angle in sputtering of alloys," *J. Vac. Sci. Technol.* **15** (2): 319–326.

Limpaecher, R. and K.R. MacKenzie, 1973: *Rev. Sci. Instrum.* **44**: 726.

Lin, I., D.C. Hinson, W.H. Cass, and R.L. Sandstrom, 1984: *Appl. Phys. Lett.* **44** (2): 185.

Mogab, C.J. and H.J. Levenstein, 1980: "Anisotropic etching of polysilicon," *J. Vac. Sci. Technol.* **17** (3): 721–730.

Muray, J.J. and I. Brodie, 1982: "Ion-assisted processes," Microscience (SRI International, Menlo Park, CA), September, pp. 1–86.

Oehrlein, G.S., G.J. Coyle, J.C. Tsang, R.M. Tromp, J.G. Clabes, and Y.H. Lee, 1986: "Plasma processing," J.Coburn, R.A. Gottscho, and D.W. Hess eds., *Mater. Res. Soc. Proc.* **68**: 367.

Perry, A.J. and R.W. Boswell, 1989: *Appl. Phys. Lett.* **55**: 148.

Schaible, P.M., W.C. Metzger, and J.P. Anderson, 1978: "Reactive ion etching of aluminum and aluminum alloys in an RF plasma containing halogen species," *J. Vac. Sci. Technol.* **15** (2): 334–337.

Schaible, P.M. and G.C. Schwartz, 1979: "Preferential lateral chemical etching in reactive ion etching of aluminum and aluminum alloys," *J. Vac. Sci. Technol.* **16** (2): 377–380.

Tokunaga, K. and D.W. Hess, 1981: "Aluminum etching in carbon tetrachloride plasmas," *J. Electrochem. Cod.* **127** (4): 928–932.

Winkler, U., 1983: "VLSI Polysilicon etching: A Comparison of different techniques," *Solid State Technology* **26** (4): 169–172.

Chapter 8

THIN FILM DEPOSITION

8.1. TECHNOLOGY TRENDS

Deposition of semiconductor thin films is accomplished by a wide variety of techniques ranging from physical processes such as evaporation and sputtering to chemical methods such as chemical vapor deposition. The properties of the films required for these applications and the film quality produced by the numerous deposition techniques will be addressed in this section.

There are three fundamental deposition methods in common use in LSI and VLSI applications:

- Evaporation
   - Resistance
   - Electron beam
   - Ionized cluster
- Sputtering
   - dc or rf diode
   - dc or rf magnetron
   - Ion beam

- Chemical Vapor Deposition (CVD)
  — Atmospheric pressure (APCVD)
  — Low pressure (LPCVD)
  — Plasma enhanced (PECVD)
  — Laser activated (LACVD)

8.1.1. Evaporation

Evaporation is accomplished by increasing the temperature of a material to its melting point at which time it evaporates and then condenses on a substrate as a thin film. Evaporation is carried out at low pressure (<10^{-6} Torr), which results in little residual gas incorporation in the growing film. Substrate heating further reduces gas inclusion (Holland 1956).

The temperature of the source is raised by two methods: filament or boat resistance, and electron beam. Resistance heating is satisfactory for low melting-point materials such as aluminum. Refractory metals, contained in water-cooled crucibles, can be melted by an electron beam. E-beam evaporation is shown schematically in Figure 8.1 (Brodie and Muray 1982).

Evaporation was most widely used for the deposition of aluminum. However, the increasing complexity of circuits has mandated the use of new metals and alloys for interconnects. Film stoichiometry is difficult to maintain when an alloy is evaporated because of the differences in melting points and thus the evaporation rates of the constituents. The wide usage of Al-Si-Cu and silicides has caused decreased usage of evaporation as a method of deposition, although silicides are commonly co-evaporated from a metal and a silicon source. The major usage is currently for GaAs devices, where lift-off techniques are facilitated by a using evaporation.

Additional disadvantages include:

- Poor step coverage
- Radiation damage from the E-gun
- Poor control of deposition rate and film thickness

Ionized cluster deposition is a method developed at Kyoto University and commercialized by Eaton. Vaporized-metal clusters are formed by the adiabatic expansion of metal vapor into a vacuum region without using inert gas as a carrier. The clusters, which consist of 500 to 2000 loosely coupled atoms, are obtained by ejecting the vapor of solid materials through the nozzle of a heated crucible into a high-vacuum region. The clusters are ionized by electron bombardment and accelerated to the substrate surface. Thin films of metals and dielectrics can be deposited. This technique is illustrated in Figure 8.2 (Ina 1988).

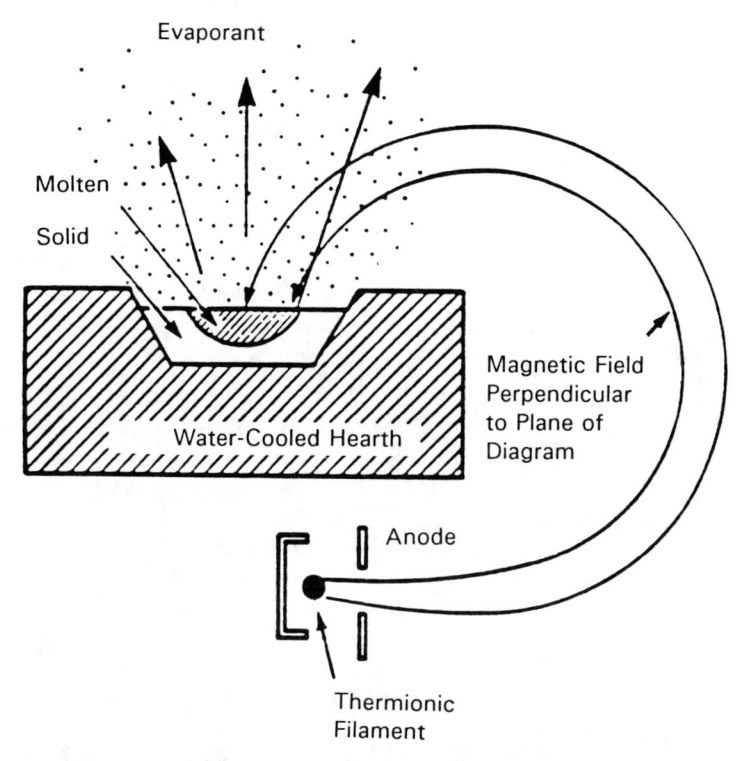

Figure 8.1. Schematic of electron beam evaporation method.

8.1.2. Sputtering

Sputtering is a process whereby a target of the material to be deposited is bombarded by high-energy (>500 eV) argon ions that are accelerated by a dc or rf field (Chopra 1969). Conventional diode sputtering, carried out at 10^{-2} Torr, can be used to deposit metallic films (dc mode) or dielectric films (rf mode). The sputtered atoms or ions arrive at the substrate with sufficient kinetic energy (20 eV) to cause mobility along the surface before it becomes thermalized. This can improve step coverage without the need for in-situ substrate heating that is used in evaporation.

Magnetron sputtering is an evolutionary process in the diode sputtering technique. Electrons are trapped near the surface of the target by perpendicular electric and magnetic fields. These electrons collide with the argon atoms

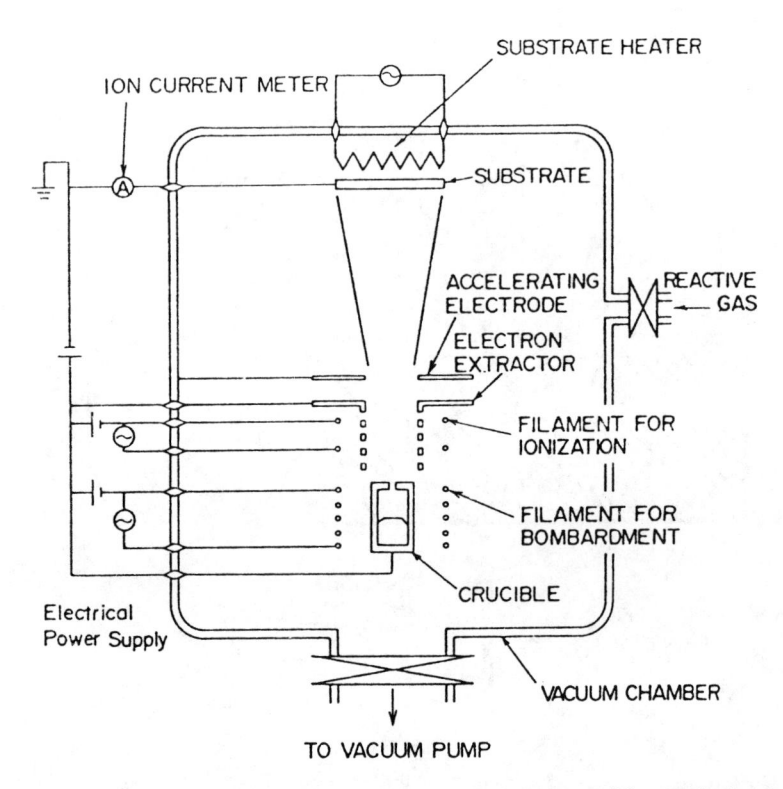

Figure 8.2. Schematic of ionized cluster beam deposition technique.

to form a plasma that has an ion-current density an order of magnitude greater than that in conventional diode sputtering. The argon ions are accelerated to the target, creating a high rate of sputtering. The magnetron target (cathode) enables sputtering to be carried out at lower pressures and temperatures, and higher deposition rates than in conventional sputtering.

The most recent designs improve the quality of the plasma by placing samarium cobalt magnets in auxiliary rings around the central magnet. Illustrated in Figure 8.3 is a diagram of a planar magnetron sputtering source.

Ion beam deposition uses an ion beam as a source of ionized particles to bombard a target and deposit a thin film (Schmidt et al. 1972). A major technological advantage over diode sputtering is that the target and substrate are not in a plasma environment. As a result, several factors that can affect thin film properties—angle of deposition, relative position of target and substrate, temperature, etc.—can be varied independently. This method is shown in Figure 8.4.

THIN FILM DEPOSITION

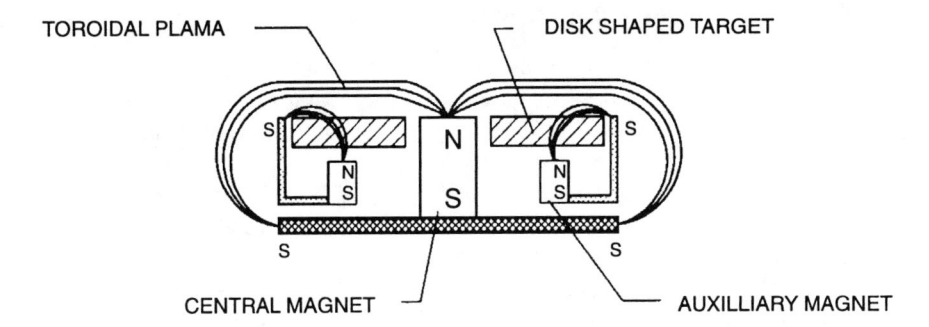

Figure 8.3. Diagram of planar magnetron sputtering source.

Sputtering is being used in the semiconductor manufacturing for the deposition of all constituents of three-dimensional interconnect systems.
Disadvantages of this method include:

- Contamination from:
  — Trace contaminants in targets
  — Impurities in the sputter gas
  — Polymer deposition
- Radiation or ion damage
- Need for high temperatures to assure adequate step coverage

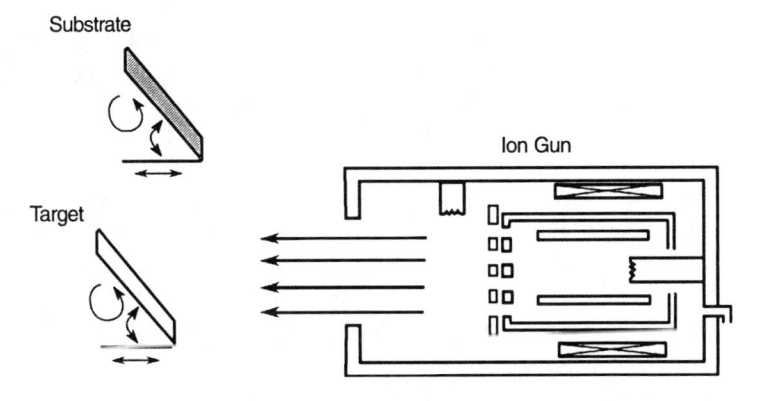

Figure 8.4. Schematic of ion beam deposition technique.

8.1.3. Chemical Vapor Deposition

Chemical vapor deposition, as the name implies, is a technique that utilizes energy to decompose a liquid or gaseous chemical compound into its constituent elements (Powell *et al.* 1966). For example, silane (SiH_4) is decomposed to form a Si film or, by means of a gas-phase chemical reaction, silicon nitride or oxide can be deposited. A number of commercially available systems supply energy to the compound in a number of different ways and thus initiate a chemical reaction. Heat is the more established method, applied to the reactor in which the wafers are maintained at atmospheric pressure (APCVD) or reduced pressure (LPCVD).

APCVD was nearly abandoned by the semiconductor community a few years ago in favor of more advanced techniques. Three common problems prevailed:

- Poor delivery control of the chemical vapor to the silicon surface
- Poor or erratic removal of reaction by-products
- Workroom interference

Use of LPCVD minimized the problem, since the vacuum-based equipment increased the mean free path, removed by-products, and incorporated a hermetic enclosure.

Advanced APCVD has returned, due to equipment developments at Watkins-Johnson. The equipment also eliminates the above problems by incorporating:

- An advanced injector
- Self-cleaning flowmeter
- Reactor designs more tolerant to variations in workroom conditions

This system, illustrated in Figure 8.5, has been used to deposit BPSG, BSG, and Tungsten. Triple-layer polycides of silicon, tungsten silicide, and silicon have been successfully made in a single in-line process. With the Watkins–Johnson Model 9000, wafers are moved on a stainless steel belt through a multi-zoned tunnel. A belt/robot combination moves wafers between cassettes and the chamber. Up to 50, 150 mm wafer can be processed at a time.

LPCVD systems were developed initially from oxidation/diffusion furnaces with the addition of gas-injection plumbing at one end and a vacuum pump at the other. These are commonly referred to a hot-wall reactors, and are illustrated in Figure 8.6. Limitations of this type of system are:

- Need for frequent quartzware changes
- Complicated quartzware for good uniformity

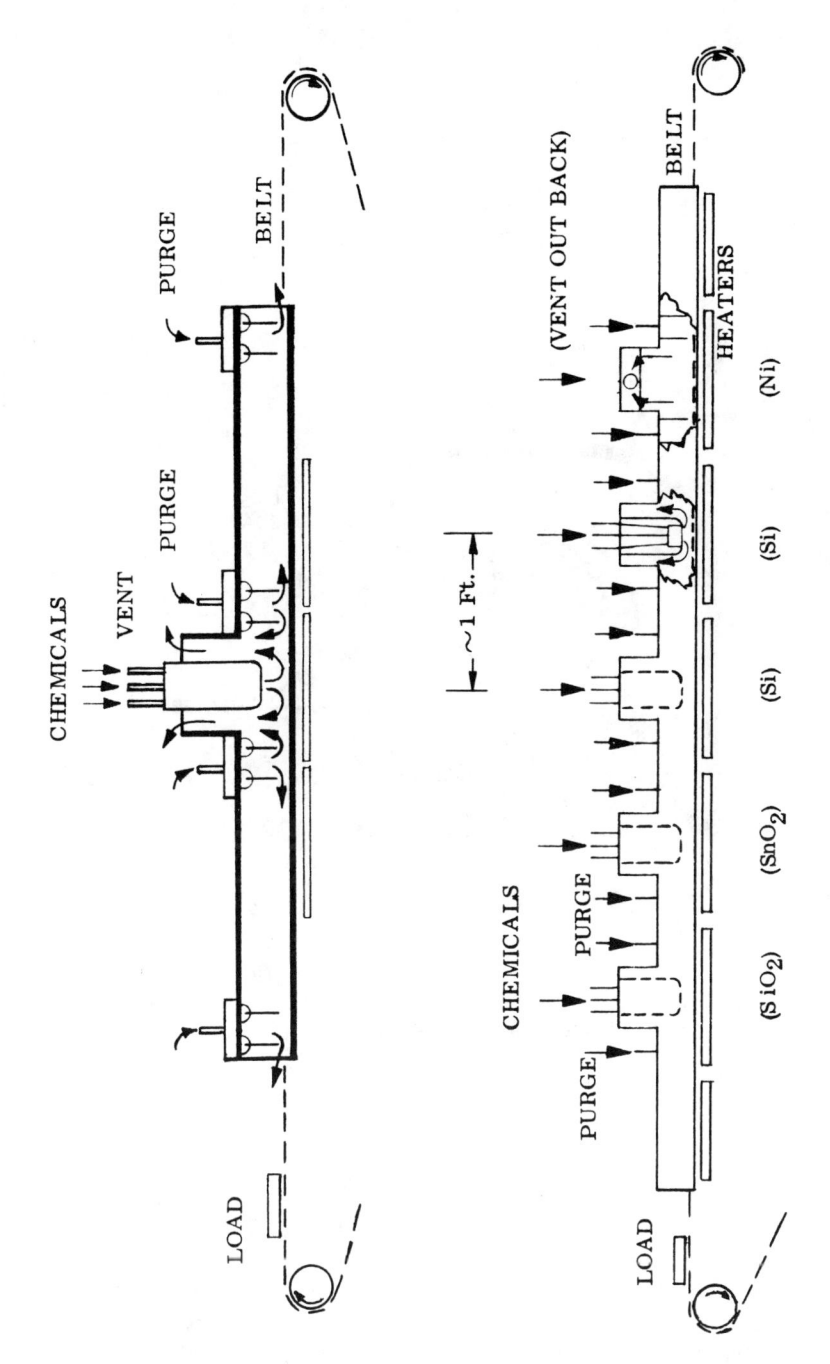

Figure 8.5. Schematic of advanced APCVD.

180 **THIN FILM DEPOSITION**

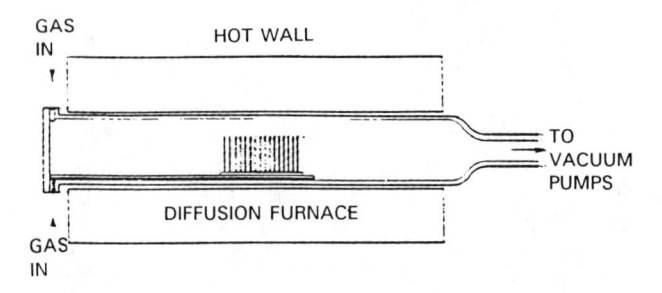

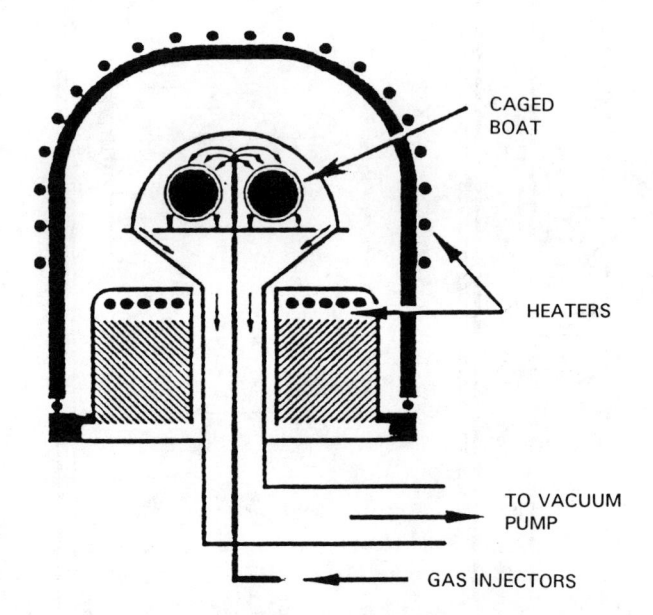

Figure 8.6. **Schematic of hot-wall LPCVD reactors.**

- Difficulty of automating wafer loading/unloading
- Need to preheat reactive gases for some films

These issues have been addressed in an isothermal design, as shown in Figure 8.6.

Cold-wall reactors represent an alternative LPCVD design. In a cold-wall design, the heated area is limited to the wafer or the platen holding the wafer. This type of reactor has been optimized for deposition of films requir-

THIN FILM DEPOSITION

ing highly-reactive gases such as selective and blanket tungsten, discussed in Section 8.2.3. In these processes, by-products in the processing of selective and blanket tungsten have been found to reduce the selectivity of tungsten deposition. In a hot-wall reactor, tungsten deposits on the walls of the reactor as well as the substrate. These reaction by-products, by reducing the selectivity, limit the film thicknesses to 50 to 300 nm, compared to 3 μm in a cold-wall reactor. This type of reactor is illustrated in Figure 8.7.

Reduced processing temperatures, necessary to prevent diffusion of shallow junctions and interdiffusion of metals in VLSI applications, have been the driving force behind the use of lower temperature methods. Plasma-enhanced deposition (PECVD) uses an rf glow discharge to generate highly reactive species.

Additional effects of high processing temperatures include:

- Induced grain growth that can cause surface roughness
- Thermal stress that can cause film peeling
- Chemical and metallurgical interactions with other materials or gases
- High adhesion due to chemical bonding with the substrate

There are two types of PECVD systems: hot-wall longitudinal and parallel plate. These are illustrated in Figure 8.8.

Laser activated CVD (LACVD) uses an excimer laser beam focused onto a cell containing the substrate and a gaseous reactant as the source of energy. The laser radiation is absorbed by the reactants that are then deposited on the substrate. The technique has been used to deposit films of silicon dioxide and silicon oxynitride. Commercial equipment was developed by Photolytics, a unit of General Signal, but abandoned by the company.

The distinction between LPCVD and PECVD has begun to blur in recent years, as LPCVD reactors have combined plasma enhancement for in-situ cleaning and substrate preparation, while PECVD systems had extended their operating range. An advanced single-wafer CVD reactor that can operate with and without plasma enhancement over the temperature range of 100° to 950°C has been developed by Varian.

Future systems will further combine technologies. Cluster tools are capable of multi-process reactions, incorporating etching, CVD, and PVD. Systems will become more flexible, with the following features:

- More sophisticated automation
- Capability of volume production and low-volume runs
- Ability to change wafer size readily

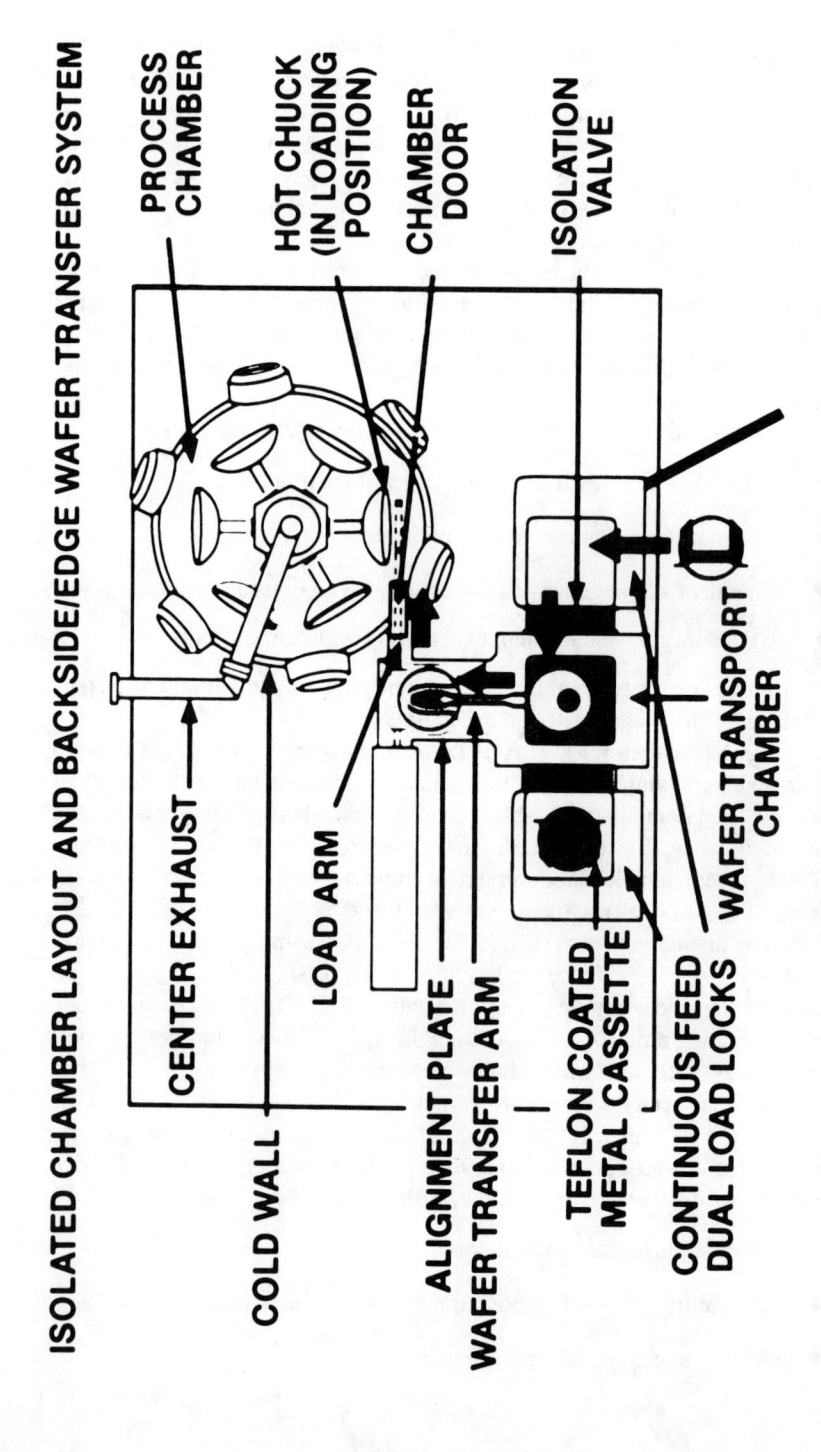

Figure 8.7. Schematic of cold-wall LPCVD reactor.

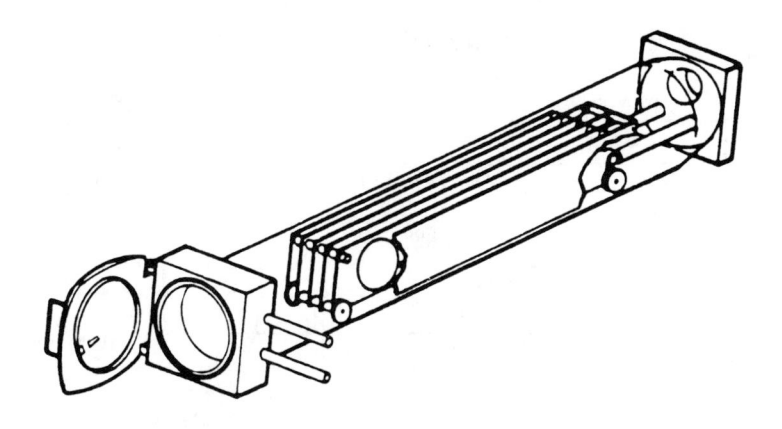

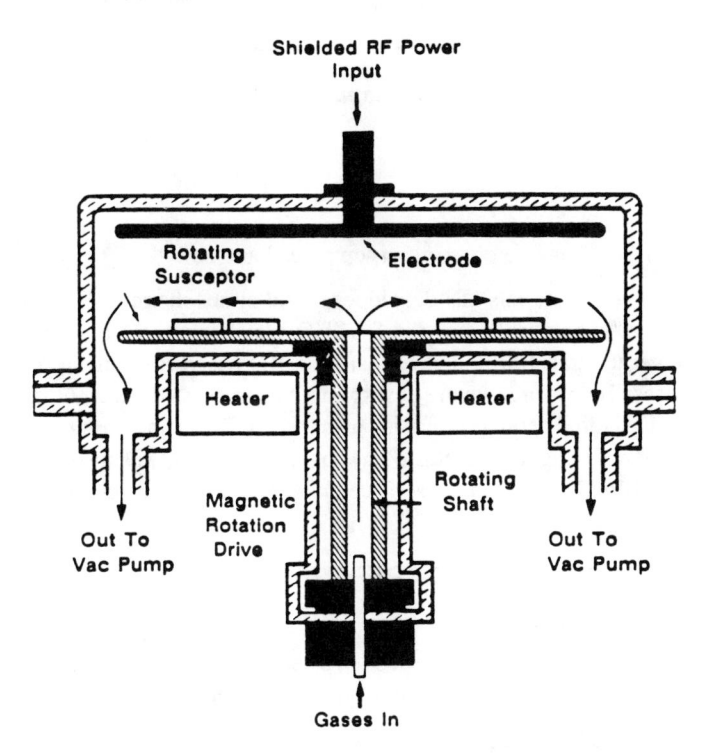

Figure 8.8. Schematic of PECVD techniques.

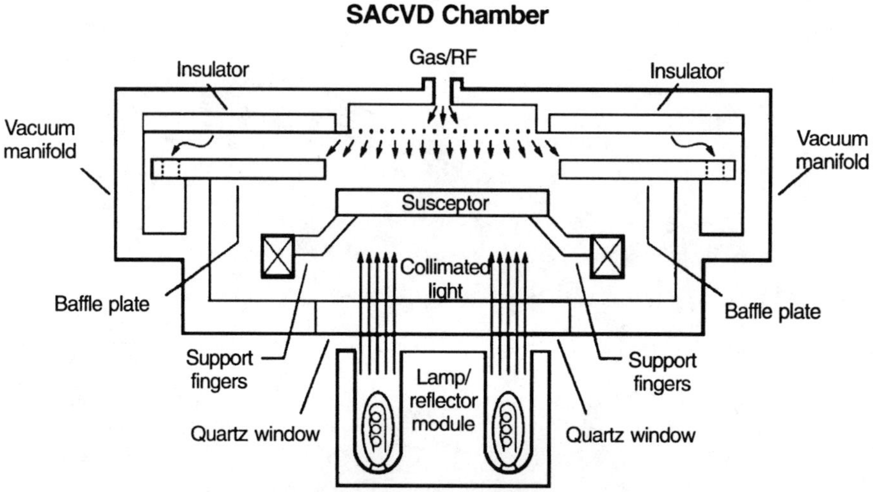

Figure 8.9. Schematic of SACVD technique.

- Particulate free
- Ion-beam or CF_4 cleaning in separate chamber

Applied Materials has developed a sub-atmospheric CVD (SACVD) process, as shown in Figure 8.9. SACVD provides a more conformal step coverage and gap-filling ability than PECVD-based inter-metal dielectric processes. To lessen moisture absorption, the process offers an *in situ* composite deposition sequence that isolates the SACVD film from exposure to atmosphere during manufacturing by covering the SACVD layer with a thin PECVD film. Deposition takes place under vacuum in one chamber.

8.2. INTERCONNECTION DEPOSITION

The most significant issue impacting VLSI technology is that of multilevel metallization. Metallic films deposited by sputtering or evaporation are used in semiconductor manufacturing in a number of applications (Sachdev and Castellano 1985).

- *Ohmic and Schottky Contacts.* Platinum silicide (PtSi) is the most commonly used material for ohmic contacts to n+ and p– Si and Schottky contacts in n-type Si in shallow (< 1 μm) bipolar devices (Berger *et al.* 1989). For deep junctions (> 1 μm), aluminum (Al) and its alloys are still used. However, for MOS devices, the most common materials are Al–Si and Al–Si–Cu.

- *Selectively Deposited Tungsten (W).* Using CVD, W deposits only in the contact vias to Si and serves as its own barrier metal prior to Al–Si deposition. This is a more reliable and simple contact than films produced by sputtered PtSi/TiW/Al-Si-Cu.

- *Barrier Metals.* Sputtered and CVD TiW is the most widely used barrier metal between Ohmic/Schottky contacts to Si and the first level of interconnect. Without this barrier, Al often spikes through the PtSi, resulting in unreliable contacts and leaky junctions. A more recent development, as discussed previously, is the use of selectively deposited tungsten.

- *Source/Drain (S/D) and Interconnect Layers in Bulk Si.* Arsenic (As) is the most commonly used n+ S/D and interconnect in bulk Si. This is diffused or implanted into regions in Si and gives resistivities of 12 to 20 ohm/sq. Boron (B) is used for p+ S/D and interconnect but has resistivities of 25 to 50 ohm/sq. New technologies have reduced these levels to 5 ohm/sq by strapping the n+ and p+ regions with sputtered or CVD refractory metal silicides such as $TiSi_2$ and $MoSi_2$.

- *Subsequent Interconnect Levels.* Sputtered Al–Si–Cu alloys are the most widely used films for interconnect materials. Si is added to reduce spiking of Al into Si while Cu is added to reduce hillock formation and increase the electromigration threshold of the Al. Problems arise in attempts to dry process films with Cu levels greater than 2%. Cu is generally not added to the top level metallization because of problems associated with wire bonding. New technology in VLSI fabrication utilizes refractory metal silicides for interconnects. Refractory metals have resistivities of <5 ohm/sq and are easier to deposit. However, they are low-temperature metallizations that require process temperatures of <500°C to avoid oxidation of the film.

8.2.1. Aluminum and Aluminum Alloys

Aluminum thin films can be deposited by evaporation, sputtering, and CVD. Aluminum alloys have gained universal usage in LSI and VLSI technologies for improved performance; Al–Cu improves electromigration resistance and Al–Si reduces hillock formation (Learn 1976). In the evaporation of these alloys, it has been difficult to maintain film composition. In addition, coverage over steps is poor, causing electrical shorts or high resistance, even when planetary motion is used.

With sputtering, however, no change in processing or fixturing is required. Films can be deposited from a sintered target alloy or co-sputtered from separate targets. Sputtering offers improved stoichiometry over evaporation and, since the sputtered atoms arrive at the wafer with some kinetic energy, it also offers improved adhesion and step coverage.

DC magnetron sputtering of aluminum alloys gives conformal coverage over sloped and vertical topographies. Over sloped walls (60°–70°) the coverage is 70% of the horizontal thickness but only 25% over a 1.5 μm vertical step (Hartsough and Denison 1979). Application of a –25 volt rf bias improves the coverage to more than 50%. No heating is necessary. It is known that copper is mobile at high temperatures and tends to agglomerate within the film to form islands that are difficult to remove by dry processing (Denison and Hartsough 1980).

Studies have shown that aluminum alloy films prepared by dc magnetron co-sputtering exhibit similar properties such as resistivity, reflectance, and composition uniformity to those films deposited from a single alloy target. An advantage of co-sputtering is that the power to each target can be varied if the composition of the film requires change (Nowicki et al. 1982).

Thin films of aluminum have been deposited in a hot-wall LPCVD reactor (Cooke et al. 1982). Tri-isobutyl aluminum (TIBA) decomposes above 260°C to give aluminum, hydrogen, and isobutylene. ASM is the sole manufacturer of a deposition system for CVD aluminum. An Al–Si alloy can be formed by heating the deposited aluminum film in-situ in the presence of silane. Silane decomposes on the fresh aluminum surface to form silicon that is taken into solid solution at 250°C to levels of about 0.5%. The resistivity of pure Al films is within 10% of bulk Al. Step coverage over a 2 μm vertical step is nearly 100%.

CVD aluminum can also be alloyed with copper to improve electromigration resistance. The conformal CVD layer is covered with sputtered Al–Cu (Flamm 1991).

TIBA is a pyrophoric liquid, however, and must be handled under inert gas. Liquid TIBA can be rendered non-pyrophoric by dilution in saturated hydrocarbons, which are also used to rinse the cold trap and evaporator to clear the non-volatile aluminum alkyl wastes.

Deposition source purity is referred to in terms of "nines"; the greater the number of nines, the greater the purity. While a material with too many impurities may adversely affect the performance of a device, it is often the type of impurity that is the source of device degradation.

Materials with alpha-emitting contaminants such as uranium and thorium affect device performance by creating soft errors that, if in the active region of a memory device, cause electrical holes (May 1979). This source of impurity is found in many materials, such as titanium tungsten, aluminum, and refractory metal silicides. Current minimum levels are 1 ppb, which will drop as smaller devices are produced. However, the current analytical equipment listed below may not be accurate below 1 ppb levels (Marx and Murphy 1990):

- Glow discharge mass spectroscopy (GDMS)
- Inductively coupled plasma/mass spectroscopy (ICP/MS)
- Neutron activation analysis (NAA)

Prices of thin film metallization materials doubles for every tenfold increase in purity. Every supplier offers the material in a variety of purities to satisfy the requirements of the user. However, as discussed above, it is the specific contamination that can have deleterious effects on a device, and further processing can degrade the purity of the films.

The key goals for future deposition procedures of aluminum are to increase the electromigration resistance, improve step coverage and uniformity, and optimize the film properties.

Advances in sputtering equipment has resulted in complex target geometries with uniform crystallographic orientation (texture) very difficult, if not impossible to maintain. Natural variations in the solidification of as-cast starting materials is associated with a variability of target grain structure and texture, making it necessary to develop new methods to reproducibly control the microstructure of the target. MRC has developed a proprietary process to lower grain size below 500 μm, and studies on these targets show a direct correlation between grain size and target performance. The reduced grain size also keeps the size of intermetallic precipitates below 10 μm. Studies have shown that precipitates larger than 10 μm lead to arcing and decreased chip yields (Marx and Murphy 1990).

Six nines purity aluminum will become the standard in the 1990s, requiring improved refining processes. Si and Cu content has been decreasing to 1 percent or less. As contact size decreases further, users will move to pure aluminum (or Al–Cu alloys for electromigration resistance) over TiW or TiN barriers. New materials such as Al/Si/Pd will be utilized for enhanced electromigration resistance and etchability (Onuki *et al.* 1988).

8.2.2. Polysilicon and Silicides

Polysilicon films for IC devices have been deposited by evaporation, sputtering, LPCVD, and PECVD. For the past few years, most polysilicon films for MOS gate electrodes and interconnects have been deposited by LPCVD. The reaction involves the thermal decomposition of silane (SiH_4) near 625°C.

The pyrolysis reaction is very temperature-sensitive, resulting in significant thickness variations. In addition, phosphorous is not easily incorporated during deposition so that post-deposition doping is required (Rosler 1977).

PECVD has been studied as a method for depositing doped and undoped polysilicon films. The process is less temperature-sensitive and considerable quantities of dopant can be incorporated during deposition (Kamins and Chiang 1982).

During the past 15 years, the replacement materials for aluminum and polysilicon have centered around refractory metal silicides. Silicides of tungsten, molybdenum, and tantalum have reasonably good compatibility with IC fabrication technology (Mochizuki *et al.* 1977).

The silicides have been deposited by electron-beam co-evaporation, CVD, and sputtering. Sputtering methodologies include co-sputtering from a metal and silicon target (Murarka and Frazer 1980), sputtering from a silicide target (Crowder and Zirinsky 1979), or reactive sputtering from a metal target in a silane atmosphere. Co-evaporation has produced films with marginal step coverage. Significant shrinkage of the films during annealing has also resulted in poor adhesion (Murarka 1981). This technique is rarely used in IC fabrication.

Co-sputtering from metal and silicon targets has produced films with better step coverage, particularly with an applied bias. Silicide composition can be adjusted by varying the power to each target. This affects film resistivity; resistivity decreases with increasing metal content (Campbell *et al.* 1982). Post-deposition heat treatments reduce the film resistivities to acceptable IC device levels. However, metal-rich silicide films tend to crack and peel on heat treatment (1000°C), which is indicative of severe stress levels. It is suspected that this is due to the significant levels of argon as inclusions in the films with resultant shrinkage during the annealing process (Whittman, 1979).

Sputtered films from silicide targets generally have high resistivities attributed to oxygen and carbon contamination. It is difficult to maintain consistent film composition in reactive sputtering.

PECVD has been successfully used for depositing silicides. Tungsten silicide has been deposited by a reaction between tungsten hexafluoride (WF_6) and SiH_4 (Akimoto and Watanabe 1981). The Si:W ratio, and hence structural and electrical properties of the films, can be adjusted by controlling the ratio of the two gases. As-deposited films have resistivities dependent on metal concentrations; high Si ratios have resistivities similar to polysilicon that decreases with decreasing Si. These films have similar resistivities to sputtered films (Miller 1980a), 350 microhm-cm and 500 microhm-cm respectively for an Si:W ratio of 1.2. Tungsten silicide has been deposited in a cold-wall LPCVD reactor at temperatures between 350°C and 450°C (Brors *et al.* 1983). The Si:W ratios and resistivities were found to be a function of

Table 8.1. **Deposition techniques for silicide formation.**

Consideration	Co-sputtered	Sputtered hot pressed target	Sputtered cold vacuum pressed target	CVD
Silicide can be formed on any material	Yes	Yes	Yes	If substrate accepts separate temperature
Control of metal-to-silicon ratio	Yes	Yes	Yes	Difficult
Purity of silicide	Good	Poor	Excellent	Acceptable
Sintering Environment Sensitivity	Not very sensitive	Same	Same	Same
After sintering surface is	Smooth	Smooth	Smooth	Rough
Possibility of selection etch	Yes	No	No	No
Deposit metal/silicon Sandwich	Yes	No	No	No
Back sputtering for clean	Yes	Yes	Yes	No

gas ratios and temperature. Step coverage on vertical steps was 75% of the horizontal film thickness (Murarka, 1983).

Illustrated in Table 8.1 is a comparison of the properties of silicide films deposited by co-sputtering, sputtering from hot pressed or cold vacuum pressed targets, and CVD.

Incorporation of gases into a growing film gives rise to increased resistivity levels. Oxygen and oxygen-carbon contamination in TiW and silicides is an important issue; oxygen contamination is thought to be a source of particulate generation. Although the addition of oxygen is required in the fabrication of barriers, precise control of the oxygen content must be maintained. Current levels of oxygen in TiW are 1,000 ppm. Newer devices will require lower levels, and concomitant material processing and refining methods.

Lower alkali metal impurities in barrier metals such as TiW is a goal in processing, as lower levels of these mobile ions results in greater control over transient electrical flow.

Other sources of contamination in the film are residual gases, the levels of which are a function of the deposition pressure, coefficient of adsorption, sticking coefficient and chemical activity of the gas. Desorption of gases from the chamber walls, as well as vacuum leaks, are additional sources of residual contaminants. High operating temperatures make it a necessity to reduce oxygen and metallic impurities in Ti. When used in contacts and barriers such as the Si/Ti/TiN/Al metallization scheme, high-purity materials are necessary to produce controlled and repeatable contact resistance while maintaining barrier performance (Ting 1982) (Tsend and Wu 1986). Currently, five nines Ti purity is standard. Alternative refining processes are being developed to improve this to six nines.

8.2.3. Refractory Metals

In the last several years, refractory metals, particularly tungsten and molybdenum, have been studied as alternative materials to aluminum, polysilicon, and silicide interconnects. The major disadvantage of W and Mo is that their oxides are volatile at temperatures above 400°C (Zirinsky *et al.* 1978). Thus, these metals can only be used when all high-temperature steps in oxidizing ambients are completed, and processing in inert ambients remains. Even after films are deposited, care must be taken to avoid oxidation during handling of wafers after heat treatment. Use of transition metals will require substantial modifications in the processing steps in IC fabrication to reduce oxidation.

Thin films of W and Mo have been deposited by electron-beam evaporation (Sinha *et al.* 1973), sputtering (Shah 1979), LPCVD (Miller and Beinglass 1980b), and PECVD (Chu *et al.* 1982). In general, films of W deposited at high temperatures (hot and cold-wall LPCVD at 400°C to 800°C) have the lowest as-deposited sheet resistivities while low-temperature methods (evaporation at 200°C and PECVD at 300°C) show the highest (Tang *et al.* 1983). Post-deposition annealing significantly reduces resistivity.

Deposition in pure WF_6 has been found unsuitable since etching is favored over deposition. Decomposition of WF_6 results in formation of F ions that etch the growing W film. Addition of hydrogen suppresses etching by scavenging the fluorine ions (Chu *et al.* 1982).

Mo films have been deposited from MoF_6 and H_2 by PECVD (Chu *et al.* 1982) and $Mo(CO)_6$ by sputtering (Oduyama 1982). PECVD films contain substantial quantities of fluorine, resulting in high resistivities. MoF_6 is also an aggressive etchant for silicon and silicon dioxide. Sputtered films contain appreciable quantities of carbon, chromium, and tungsten contaminants, but sputtering is now the choice process for molybdenum gate formation.

Tungsten as an interconnect metallization has a number of advantages over aluminum alloys or polysilicon. Although the resistivity of W is high (5.3 microhm-cm) compared to aluminum (2.7 microhm-cm) or aluminum-copper (3.2 microhm-cm), it is the thinning of the deposit at steps in VLSI

Thin Film Deposition

topographies that is the major limitation in device performance. The thinning increases the interconnect resistance significantly. CVD tungsten can be deposited conformally over a step so that little thinning occurs compared to evaporated or sputtered aluminum (Sachdev and Castellano 1985).

A property of CVD tungsten is that it does not adhere strongly to silicon dioxide. Selective tungsten deposition occurs via the reaction of WF_6 and H_2 at approximately 300°C with an H_2 partial pressure of <1 torr. The proposed mechanism is the reduction of WF_6 by the silicon surface to produce a self-limited tungsten film to about 150 angstroms. Silicon dioxide or other dielectrics do not reduce the WF_4. Hydrogen is then dissociated on the W film, causing further decomposition of WF_6 (Singer 1990).

Selective tungsten deposition has been demonstrated on silicon, polycrystalline silicon, WSi_2, $TiSi_2$, and Al.

The poor adherence of CVD tungsten on dielectrics allows it to be selectively deposited on Si gates to form a resistance barrier needed to protect shallow junctions from aluminum spiking (Miller, 1981). The tungsten barriers, 1200–1500 Å thick, are deposited through contact vias in an oxide film prior to aluminum interconnect deposition. Selective deposition saves masking and etching steps and is cheaper than sputtered contact barriers that require subsequent etching.

Materials must satisfy the following criteria to function as an IC barrier layer (Hems 1990) (Pramanik and Jain 1991):

- It must adhere to oxide and nitride films
- Aluminum must adhere to it
- It must be easily dry etched in combination with Al conductors
- There must be no excessive reaction with Al that would lead to large increases in resistance
- It must have a manufacturable deposition process
- It must provide good barrier properties with thickness <2,000 Å
- It must be characterized by low stress: 1 to 5×10^9 dynes/cm^2 compresive
- It must not cause degradation of electromigration characteristics of the interconnect
- It must have low contact resistance and resistivity
- There must be low diffusion of both silicon and metal through the barrier

Recent advances in depositing thick films of selectively deposited tungsten have extended its use to completely filling the holes. Film thicknesses of three µm have been made with aspect rations of 3:1, and at rates of 5000 Å/min.

An alternative contact barrier is a titanium tungsten film. TiW films are generally sputtered from cathodes composed of Ti(10%)W (Cunningham *et al.* 1978). This alloy is typically employed as a barrier between platinum or palladium silicide and an aluminum interconnect. The TiW and Al films are best deposited sequentially under vacuum to prevent a thermal oxide interface (Ghate *et al.* 1978).

The above process imposes difficulties with VLSI processing. The TiW film is deposited over the entire wafer prior to Al deposition. A separate processing step is required to remove the film in unwanted regions of the wafer. Either the TiW is deposited over a resist and removed from the non-contact via area by a lift-off technique or patterned along with the Al interconnect lines. This presents a dry processing problem since the etching chemistry of TiW differs from that of Al.

Titanium nitride and nitrided TiW have demonstrated superior barrier properties. The films are easily deposited by adding nitrogen to the reactor chamber during sputtering (Ting 1982).

8.3. DIELECTRIC DEPOSITION

8.3.1. Silicon Dioxide

Dielectric films used as an insulation between conductive regions of an IC must meet several important requirements to be effective:

- High purity
- Low pinhole and particulate density
- Low intrinsic stress
- High breakdown voltage
- Good adhesion
- Conformal coverage

SiO_2 meets these requirements and is used in IC fabrication. Phosphorous-doped oxide (PSG) is preferred over undoped glass, largely because of its ability to reflow at 1000°C, to reduce topography variations and to taper contact via profiles (Vossen, 1974). In addition, PSG has improved gettering and resistance to sodium penetration compared to undoped glass (Eldridge and Kerr 1971).

The current trend in low-temperature processing has generated interest in borophosphorsilicate glass (BPSG) since it can reflow at lower temperatures, is deposited with lower stress, and is a more effective passivation layer than is PSG (Kern Schnable 1982).

BPSG, as a passivation layer on bipolar and MOS devices prior to plastic encapsulation or hermetic ceramic packaging, meets the following requisites:

- Alkali barrier and gettering capabilities
- Corrosion and scratch protection of metal films
- Prevention of shorts caused by loose conductive particles
- Reduced surface charge buildup

Most of the current IC processing technologies use CVD as the method of choice for deposition. APCVD suffers some drawbacks compared to LPCVD, mainly in thickness uniformity, defect density, and throughput (Rosler and Engle 1981).

With this trend in low-temperature processing, PECVD has gained increased acceptance in depositing BPSG (Hills *et al.* 1990). The advantages over other CVD methods include (Avigal 1983):

- Deposition at <400°C (then densified at 800°C)
- Lower defect density
- Improved step coverage
- Lower particulate density
- Reduced hillock formation of Al underlayer

Microwave enhanced deposition has shown reduced hydrogen content in the oxide films. MTI's afterglow reactor has been used to deposit films at 3000 Å/min with a hydrogen content of only 1% compared to 6% for PECVD. Electron cyclotron resonance (ECR) reactors have deposited films with a similar hydrogen content of 1%. Also silicon nitride films had no hydrogen detected, compared to as much as 20% for PECVD. The order of magnitude lower operating pressure and greater ion flux are significant factors.

TEOS (tetraethylorthosilicate)/ozone-based BPSG have better conformality than silane-based BPSG. APCVD BPSG is currently the system of choice for reflowed glass used prior to first-level metal deposition in 1 and 4 Mbit DRAMs. LPCVD BPSG using TEOS is an area of high activity among vertical tube reactor companies such as ASM, BTU, Kokusai Electric, SVG, and TEL. These companies are offering single tube, two-step, *in situ* deposition and reflow as a BPSG integrated process solution. Low-temperature PECVD TEOS-based BPSG is already used for interlayer dielectrics. Applied Materials' Precision 5000 system is capable of pyrolytic and plasma enhanced deposition of SiO_2 from TEOS at 375°C. PECVD manufacturers such as Applied, ASM, Electrotech, and Novellus are offering the capability of thermal CVD using TEOS-ozone at pressures close to atmospheric without plasma enhancement.

8.3.2. Silicon Nitride

PECVD of silicon nitride is probably the industry standard and is used in production in most IC lines throughout the world as a final passivation layer. It meets all the requirements listed in the previous section and is a better sodium and moisture barrier than oxide is. It also offers advantages over plasma oxide as a dual-layer isolation because of moisture and ion impermeability, ease of patterning by dry processing techniques, and ease of deposition (Mar and Samuelson 1980).

Important compositional considerations of silicon nitride is the silicon-to-nitrogen ratio and the amount of hydrogen incorporated in the film. Thermally deposited films have the composition Si_3N_4 while the composition varies with plasma nitride. The refractive index of the film indicates the silicon-to-nitrogen ratio; the correct composition has an index of 2.05. Equally important is the amount of hydrogen released during subsequent processing. By adjusting process parameters, it is possible to reduce the SiH in silicon films (Harrus and van de Ven 1990a).

Hydrogen incorporation is most important for IC devices and depends on the deposition temperature and reactant gases. The average hydrogen content of a PECVD nitride film is 20% (Lanford, 1978), whereas little is incorporated in thermal nitride because of the high deposition temperatures. Nitrogen released during subsequent processing can be detrimental for certain devices. Silicon oxynitride films formed by PECVD have been reported with low permeability, lower intrinsic stress, and lower hydrogen content than PECVD nitrides at deposition temperatures of 300 to 350°C.

A direct correlation between reduced amounts of SiH bonds and improved degradation characteristics of devices has been made using low-hydrogen nitride, oxynitride, and fluorinated nitride.

Because of increased line resistance and RC delays in high speed memories, passivation schemes are evolving toward a bilayer sandwich with oxide or BPSG as the bottom layer and nitride as the top (Gootzen *et al.* 1989). The immediate advantage of oxide between metal runners is a reduced dielectric constant from 7 for nitride to 4. To eliminate voices in these narrow metal spaces, TEOS-based PECVD films (Chin and van de Ven 1988) is followed by a highly conformal nitride film (Harrus *et al.* 1990b).

Table 8.2. Comparison of film properties for PECVD passivation materials.

Film properties	SiN:H	UV-Nitride	TEOS	Oxide/PSG	Oxynitride
Chemistry	$SiH_4/NH_3/N_2$	$SiH_4/NH_3/N_2$	$TEOS/O_2/TMP$	$SiH_4/N_2O/PH_3$	$SiH_4/NH_3/N_2O$
Deposition rate	2000 Å/min.	500 Å/min.	2500 Å/min.	5000 Å/min.	1800–2500 Å/min.
Uniformity	1.5%	1.5%	1.5%	1.0%	1.5%
Stress	Controllable	Controllable	Controllable	Controllable	Controllable
	2×10^9 d/cm^2	2×10^9 d/cm^2	1.5×10^9 d/cm^2	1.5×10^9 d/cm^2	1×10^9 d/cm^2
Step coverage	Good	Good	Very good	Poor	Good
Refractive index	2.01 ± 0.02	1.95 ± 0.02	1.465 ± 0.015	1.465 ± 0.015	(1.6 to 1.9) ± 0.02
Hydrogen content	12–15 at%	≤ 2 at%	N/A	N/A	N/A
UV transparency	None	$\geq 85\%$ at 254 nm	Yes	Yes	Yes
Moisture barrier	Good	Good	Poor	Poor	Acceptable

Table 8.2 gives an overview of the properties of numerous PECVD-deposited dielectric films and passivation schemes (Harrus and van de Ven 1990a).

8.4. TECHNOLOGY COMPARISONS

8.4.1. Evaporation Versus Sputtering Versus CVD

Evaporation is commonly used for depositing aluminum and refractory metals. The high purity of the films and the inexpensive, non-complex equipment makes this method the preferred choice when film properties and device characterization permit.

Although VLSI device complexities have dictated the use of alternative deposition techniques for state-of-the art metallizations, evaporation is still a viable technique. Lift-off techniques for patterning is one area of technology that is making significant use of evaporation, since it takes advantage of the poor conformal coverage of evaporated films. The lift-off of a resist and overlayer film is facilitated if this film is not continuous over the resist profile. Lift-off is one of predominant methods of patterning used in GaAs technology. It is also used in planarization of silicon devices.

Nevertheless, the use of evaporation will be supplanted by CVD. Use of line-of-sight deposition to fill contacts or channels will also find decreased usage because of the non-selectivity of the process compared to CVD.

Sputtering is used predominantly for the deposition of aluminum alloys, refractory metals, and silicides. As stated previously, this method offers the advantages of compositional control and conformal coverage compared to evaporation. Sputtering will continue to be used as the dominant technique for aluminum alloy deposition in VLSI devices.

However, silicides and refractory metals will become the dominant metallization in VLSI devices, and sputtering will lose market share to CVD for these materials.

Use of CVD has been growing with the development of low-temperature processes (PECVD and LACVD), the increasing usage of refractory metals (salicides and silicides), novel applications (tungsten plugs), and advanced equipment. This technology will become the dominant force in the deposition industry.

8.4.2. Single Versus Batch Processing

Deposition equipment in Section 8.1 was described on the bases of type of equipment. System descriptives can further be based on chamber configuration and lot size, as described below:

- *Multi-chamber, single wafer per chamber* systems have recently been developed to overcome the low throughput of deposition a film on one wafer at time. One such system is the Applied Materials' Precision 5000, built around a pentode-shaped vacuum loadlock containing a wafer buffer and a frog-type wafer loader. The system accomodates up to four process chambers and a cassette load station. Each chamber requires its own set of controllers, pumps, heaters, and gate valves.

- *Single chamber, small batch* systems are designed to accommodate a small batch of up to 15 wafers, depending on wafer diameter. One such system is Electrotech's Model ND 6210, a PECVD system whereby wafers are placed on a round susceptor in the process chamber. This system was the first to incorporate a fully automated cassette-to-cassette load/unload.

Nearly all sputtering and evaporation systems can be classified in this category.

- *Single chamber, large batch* systems include several versions:
  — A thermal deposition system based on a diffusion furnace, such as Advanced Crystal Sciences' Model Falcon IV, whereby 100, 200 mm wafers are loaded vertically in slotted quartz boats.
  — A PECVD system based on a diffusion furnace, such as the ASM plasma system, where wafers are loaded onto arrays of heavy graphite susceptors.
  — A thermal system with a bell jar configuration, such as the Anicon (Silicon Valley Group) Model 1200 hot-wall reactor, where boats are loaded side-by-side rather than end-to-end for up to 150 wafers.
  — Focus's Model F1000, a large batch system with a non-loadlocked vertical heater block with none wafers mounted on each side. A cover plate has nine cupped depressions that fit over each wafer.
  —Anelva's Model 1015 DC magnetron sputtering system, capable of holding 100, 150 mm wafers.

- *Continuous process, large batch*, with two examples:
  — The Watkins-Johnson Model WJ-999, whereby wafers are moved through a multi-zoned tunnel over a stainless steel belt.
  — Novellus Systems' Concept One in which a small round batch-type chamber can hold seven wafers, each resting under a separate dispersion head. Wafers are sequentially moved from head to head, each receiving one-seventh of its total deposition at each station. Up to 75 wafers can be processed between cycling of the loadlock.

Advantages of single wafer reactors include:

- Fewer wafers can be damaged because of equipment malfunction
- Wafer-to-wafer film thickness, step coverage uniformity grain size, and reflectivity is high
- Atmospheric water vapor can be kept out with loadlocks, used on virtually all systems
- The ease of automation in an in-line concept

There are many viable production applications for batch systems:

- In general, throughput is higher. An analysis of the above CVD systems has been performed by Novellus. Throughputs are (including deposition rate and overhead time):
  — Multi-chamber, single wafer
    - Applied Materials - 30 wafers/hr
  — Single chamber, small batch
    - Electrotech, 26 wafers/hr
  — Single chamber, large batch
    - ACS, 25 wafers/hr
    - ASM, 28 wafers/hr
    - Anicon, 33 wafers/hr
    - Focus, 38 wafers/hr
  — Continuous process, large batch
    - Watkins–Johnson, 43 wafers/hr
    - Novellus, 55 wafers/hr
- Systems are less expensive and easier to maintain, resulting in lower production costs. The analysis by Novellus also covered cost of operation:
  — Multi-chamber, single wafer
    - Applied Material,- $227,428
  — Single chamber, small batch
    - Electrotech, $906,816
  — Single chamber, large batch
    - ACS, $482,672
    - ASM, $645,900
    - Anicon, $571,635
    - Focus, $444,466
  — Continuous process, large batch
    - Watkins–Johnson, $300,056
    - Novellus, $227,428
- Co-deposition of materials such as silicides from multiple targets is easier to accomplish

REFERENCES

Akimoto, K. and K. Watanabe, 1981: "Formation of W_xSi_{1-x} by plasma chemical vapor deposition," *Appl. Phys. Lett.* **39** (5): 445–447.

Avigal, I., 1983: "Inter-metal dielectric and passivation-related properties of plasma BPSG," *Solid State Technology* **26** (10): 217–224.

Berger, S., et al., 1989: "The structure and composition of contacts made of Al(2% Cu)-TiW films on amorphized (100) silicon," *Thin Solid Films* **176**: 131.

Brodie, I. and J. J. Muray, 1982: *The Physics of Microfabrication*, Plenum Press, New York.

Brors, D.L., J.A. Fair, K.A. Monning, and K.C. Saraswat, 1983: "Properties of low pressure CVD tungsten silicide as related to IC process requirements," *Solid State Technology* **26** (4): 183–186.

Campbell, D.R., S. Mader, and W.K. Chu, 1982: "Effects of grain boundaries on the resistivity of co-sputtered WSi_2 films," *Thin Solid Films* **93**: 341–346.

Chin, B.L., and E.P. van de Ven, 1988: *Solid State Technology* **31** (4): 110.

Chopra, K.L., 1969: *Thin Film Phenomena*, McGraw-Hill, New York.

Chu, J.K., C.C. Tang, and D.W. Hess, 1982: "Plasma-enhanced chemical vapor deposition of tungsten films," *Appl. Phys. Lett.* **41** (1): 75–77.

Cooke, M.J., R.A. Heinecke, and R.C. Stern, 1982: "LPCVD of aluminum and Al-Si alloys for semiconductor metallization," *Solid State Technology* **25** (12): 62–65.

Crowder, B.L. and S. Zirinsky, 1979: "1 μm MOSFET VLSI technology, VII. Metal silicide interconnection technology—a future perspective," *IEEE J. Solid State Circuits* **SC-14** (2): 291–293.

Cunningham, J.A., et al., 1978: *IEEE Trans. Reliability* **19**: 182.

Dension, D.R. and L.D. Hartsough, 1980: "Copper distribution in sputtered Al/Cu films," *J. Vac. Sci. Technol.* **17** (1): 388–392.

Eldridge, J.M. and D.R. Kerr, 1971: "Sodium ion drift through phosphosilicate glass—SiO_2 films [MOS structures]," *J. Electrochem. Soc.* **118** (6): 986–991.

Flamm, D.L., 1991: "Aluminum surges ahead at VMIC," *Solid State Technology*, **34** (8): 47–48.

Ghate, P.B., J.C. Blair, C.R. Fuller, and G.E. McGuire, 1978: "Applications of Ti:W barrier metallization for integrated circuits," *Thin Solid Films* **53**: 117–128.

Gootzen, W.M., M. Bellersen, L. de Bruin, G. Rao, G. Rutten, and D. Yen, 1989: Proc. 6th IEEE Electr. Device Lett., 1982, pp. 90.

Harrus, A.S and E.P. van de Ven, 1990a: "New passivation schemes needed for VLSI," *Semiconductor International* **13** (6): 124–128.

Harrus, A.S., I.W. Connick, and E.P. van de Ven, 1990b: Proc. Semicon Europe, Zurich, Switzerland, March.

Hartsough, L.D. and D.R. Denison, 1979: "Aluminum-silicon sputter deposition," Technical Report 79.01, Perkin–Elmer Corp.

Hems, J., 1990: "Barrier layers: The advantages of titanium nitride," *Semiconductor International* **13** (11): 99–102.

Hills, G.W., A.S. Harrus, and M.J. Thoma, 1990: "Plasma TEOS as an intermetal

dielectric in two level metal technology," *Solid State Technology* **33** (4): 127–132.
Holland, L., 1956: *Vacuum Deposition of Thin Films*, Chapman and Hall, London.
Ina, T., 1988: "Large area film deposition by ionized cluster beam technologies," Technical Proc. Semicon West, May 24026, San Mateo, CA pp. 242–248.
Kamins, T.I. and K.L. Chiang, 1982: "Properties of plasma-enhanced CVD silicon films," *J. Electrochem. Soc.* **129** (10): 2326–2331.
Kern, W. and G.L. Schnable, 1982: "Chemically vapor-deposited borophosphosilicate glasses for silicon device applications," *RCA Review* **43** (3): 423–457.
Lanford, W.A. and M.J. Rand, 1978: "The hydrogen content of plasma-deposited silicon nitride," *J. Appl. Phys.* **49** (4): 2473–2477.
Learn, A.J., 1976: "Evolution and current status of aluminum metallization," *J. Electrochem. Soc.* **123** (6): 894–906.
Mar, K.M. and G.M. Samuelson, 1980: "Properties of plasma-enhanced CVD silicon nitride measurements and interpretations," *Solid State Technology* **23** (4): 137–142.
Marx, D.R. and R. G. Murphy, 1990: "Sputtering targets: challenges for the 1990s," *Solid State Technology* **33** (3): S11–S13.
May, T.C., 1979: *IEEE Trans.* **CHMT-2** (1): 377.
Miller, R.J., 1980a: "Resistivity and oxidation of tungsten silicide thin films," *Thin Solid Films* **72**; 427–432.
Miller, R.J. and I. Beinglass, 1980b: "Hot wall CVD tungsten for VLSI," *Solid State Technology* **23** (12): 79–82.
Miller, N.E. and R. Herring, 1981: "Thin tungsten films by CVD," 159th Meeting Electrochem. Soc., Vol. 81–1, pp. 712–713.
Mochizuki, T., K. Shibata, T. Inoue, K. Ohuchi, and M. Kashigawa, 1977: "A new MOS process using MoSi$_2$ as a gate material," *Jpm. J. Appl. Phys. Suppl.* **17**: 37–42.
Murarka, S.P. and D.B. Frazer, 1980: "Silicide formation in thin film co-sputtered (titanium + silicon) films on polycrystalline silicon and SiO$_2$," *J. Appl. Phys.* **51** (1): 350–356.
Murarka, S.P., 1981: "silicides, which, why, how?," *Semiconductor Silicon* **81** (5): 551–558.
Murarka, S.P., 1983: *Silicides for VLSI Applications* (Academic Press, New York).
Nowicki, R.S., E.V. English, L.E. Gulbrandsen, A.J. Learn, and K.W. Schuette, 1982: "Dual RF diode/DC magnetron sputtered aluminum alloy films for VLSI," *Semiconductor International* **5** (3): 105–114.
Oduyama, F., 1982: "A simple technique to deposit molybdenum thin films," *Appl. Phys. A* **28** (2): 125–128.
Onuki, J., Y. Koubuchi, S. Fukada, M. Suwa, Y. Misawa, and T. Itagaki, 1988: 1988 Intnternat. Electron. Dev. Mtg. Tech. Dig., p. 454.
Powell, C.F., J.H. Oxley, and J.M. Blocher, eds., 1966: *Vapor Deposition*, Wiley & Sons, New York.
Pramanik, D. and V. Jain, 1991: "Barrier metals for ULSI: processing and reliability," *Solid State Technology* **34** (5): 97–102.

Rosler, R.S., 1977: "Low pressure CVD production processes for poly, nitride, and oxide," *Solid State Technology* **20** (4): 63–70.

Rosler, R.S. and G.M. Engle, 1981: "Plasma-enhanced CVD in a novel LPCVD-type system," *Solid State Technology* **24** (4): 172–177.

Sachdev, S. and R.N. Castellano, 1985: CVD tungsten and tungsten silicide for VLSI Applications," *Semiconductor International* **8** (5): 306–310.

Schmidt, P.H., R.N. Castellano, and E.G. Spencer, 1972: *Solid State Technology* **15** (7): 27.

Shah, P.L., 1979: "Refractory metal gate processes for VLSI applications," *IEEE Trans. Elec. Dev.* **ED-26** (4): 631–640.

Singer, P.H., 1990: "Selective deposition nears production," *Semiconductor International* **13** (3): 46–50.

Sinha, A.K., T.E. Smith, T.T. Sheng, and N.N. Axelrod, 1973: "Control of resistivity, microstructure, and stress in electron beam evaporated tungsten films," *J. Vac. Sci. Technol.* **10** (3): 436–444.

Tang C.C., J.K. Chu, and D.W. Hess, 1983: "Plasma-enhanced chemical vapor deposition of tungsten, molybdenum, and tungsten silicide films," *Solid State Technology* **26** (3): 125–128.

Ting C.T., 1982: "TiN formed by evaporation as a diffusion barrier between Al and Si," *J. Vac. Sci. Technol.* **21** (1): 14.

Tsend, H-H., and C-Y. Wu, 1986: "A new oxidation-resistant self-aligned $TiSi_2$ process," *IEEE Elec. Dev. Lett.* **ED-7** (11): 623.

Vossen, J.C., G.L. Schnable, and W. Kern, 1974: "Processes for multi-level metalization," *J. Vac. Sci. Technol.* **11** (1): 60–70.

Whittman, M., J. Roth, and J.W. Mayer, 1979: "Influence of nobel gas atoms on the epitaxial growth of implanted and sputtered amorphous silicon," *J. Electrochem. Soc.* **126**: 1247.

Zirinsky, S., W. Hammer, F. D'Heurle, and J. Baglin, 1978: "Oxidation mechanisms in WSi_2 thin films," *Appl. Phys. Lett.* **33** (1): 76–78.

Chapter 9

LASER PROCESSING

9.1. INTRODUCTION

Since its invention in 1960, there has been a proliferation of laser types, ranging in wavelength from X-rays having short wavelengths to the far-infrared long wavelength electromagnetic wavelength. Lasers are also further sub-divided into two categories: those that emit continuous radiation (CW) and those that emit their radiation in discrete pulses. Pulsed laser emission is characterized in terms of either the power at the peak of the pulse or the total energy emitted during a single pulse.

Carbon dioxide lasers, operating in the infrared, dominate as the most widely used laser source. Today's fast-axial-flow lasers are more compact than earlier generations of slow-axial-flow and transverse-mode carbon dioxide lasers, making possible robot control of laser power. Nevertheless, the enhanced pulse capabilities of slow-axial-flow lasers are advantageous when the application calls for rapid energy coupling and low heat input. In the pulsed mode, the laser is electronically modulated to emit a pulse with an enhanced peak power several times greater than the CW power level. The material being processed is rapidly vaporized, minimizing heating of the area surrounding the beam.

The Neodymium:Yttrium Aluminum Garnet (Nd:YAG) is the most widely used solid-state laser. The short wavelength generated is more readily absorbed by metal than the carbon dioxide laser. Because of their relatively low cutting speed, Nd:YAG lasers are often Q-switched to produce narrow pulse widths (typically 200 nsec) and higher pulsed power. These have an advantage in high pulse repetition rates (100 pps) compared to the Nd:Glass laser (3 pps) used for spot welding and hole-drilling applications (Elza 1985).

Excimer lasers, although more than fifteen years old, are only recently becoming a force in semiconductor manufacturing. However, this growth is rapid because these lasers emit in the ultraviolet spectral region and capable of producing peak powers of up to 10^7 watts. The energy of a photon from an ArF excimer laser at 193 nm is 6.42 eV is more than twenty times larger than for a photon from a carbon dioxide laser (Plummer 1985).

Although the original excimer was termed for homonuclear diatomics such as argon and krypton it now includes heteronuclears and polyatomics such as ArF^*, KrF^*, $KrCl^*$, XeF^*, and $XeCl^*$.

Excimer lasers describe a group of lasers in which the lasing species (excimer) is a diatomic molecule that is bound only in electronically excited states, while the electronic ground states are purely repulsive or only weakly bound. The upper state is formed by the chemical reaction after one or both of the constituents have been electronically excited or ionized in a fast, high-voltage discharge. From this upper state, the excimer undergoes laser transitions to its repulsive or rapidly dissociating ground state.

For a KrF^* laser, the upper laser level is an ionically bound state that is formed by a three-body recombination of the Kr^+ and F^- ions in the presence of a collision partner (Brau 1984). This level decays via fast spontaneous emission and collisional deactivation with a pressure dependent lifetime of 2.5 ns in the absence of stimulated processes. Thus, ionic precursors must be prepared on a time scale and with concentrations sufficient to produce 10^{23} excimers/cc/sec and a sizable population of 10^{15}/cc. These ions can be prepared by either electron beams, microwave discharges, or fast avalanche discharge.

In avalanche discharge excitation, the gas mixture contains Kr, fluorine, and a buffer gas such as He or Ne. As shown in Figure 9.1, these gases are injected into the laser chamber between two oblong electrodes 2 to 3 cm apart and exposed to a fast-pulsed, high-voltage discharge at a total pressure of 2 to 4 atmospheres. Typical current densities are 10^3 A/cm^2 and a breakdown voltage of 25 to 35 kV and an electron temperature of 1 eV in order to provide the 14 eV ionization energy of Kr.

Laser-based systems have become increasingly important as feature sizes of semiconductor devices have decreased. Shown in Table 9.1 is a syn-

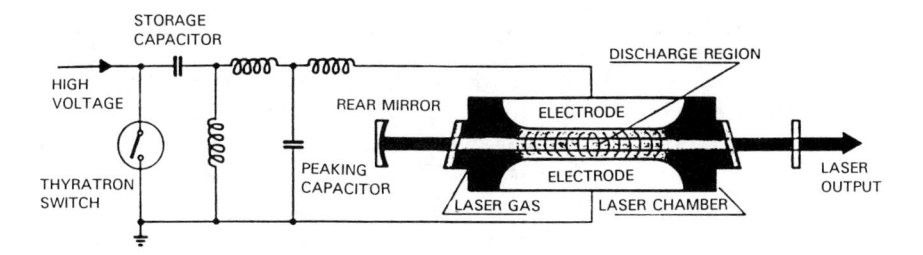

Figure 9.1. **Schematic of excimer laser.**

opsis of the major applications of lasers and the approximate timeframe of their introduction.

Lasers have been introduced into commercial systems, and are a promising technology for the development and production of advanced semiconductor devices. For example, it is probable that only a laser-based lithography system is capable of defining the 0.3 µm feature sizes required for the production of 64 Mbit dynamic random access memories in 1995.

Table 9.1. **Lasers in semiconductor processing.**

Application	Laser Type	Status and Year Introduced
Positioning/stepping	HeNe	One of earliest applications (1961)
Marking, machining,	CO_2	Dicing of wafers (1976)
scribing	Nd:YAG	Scribing and machining (1977)
	Nd:YAG	Character generation bar codes (1977)
Resistor trimming	Nd:YAG	Thin film on silicon (1975)
Link Breaking	Nd:YAG	Early work on redundant memories (1978)
Link making	Nd:YAG	Connections for IC personalization (1980)
Mask, reticle	Nd:YAG	Laboratory development (1982)
generation	Ar ion	First commercial aligner (1985)
	Ar ion	First commercial PG (1985)
	Nd:glass	First commercial X-ray (1987)
Annealing	Nd:YAG	Selective annealing (1976)
	Excimer	(1979)
Direct writing	Nd:YAG	Early work with photoresists (1980)
	Ar ion	Refined photoresists (1983)
	Excimer	UV photoresists (1985)
Material deposition	Ar ion	Laser Pantography (1983)
Nondestructive test	HeNe	Latch-up and photoluminescence (1982)
	Ar ion	(1984)
Laser inspection	Ar ion	Holographic technique (1987)
	HeNe	Wafer inspection (1987)
Laser soldering	Nd:YAG	First commercial system (1986)

9.2. Laser Reactions

Laser microchemistry can be defined as a process in which energy supplied by a finely focused laser beam drives a chemical reaction in a localized area (Oprysko et.al. 1986). The initiation of chemical reactions using high energy laser photons with energy ranging from 0.1 eV in the infrared to 6 eV in the ultraviolet has been studied for a variety of gas-solid systems (Chuang 1982a). The interaction of laser photons can occur with: (1) the gaseous species, (2) the species adsorbed on the solid surfaces, and (3) the solid substrate. In processes (1) and (2), both the electronic and vibrational excitation of the gaseous molecules can be important. In process (3), both the electronic and lattice phonon (thermal) excitation of a solid may be involved.

UV lasers dissociate the reacting molecules while IR lasers are used primarily to raise the temperature of the reacting gases. Although UV lasers have been used to successfully deposit metal and insulator films, these films have generally not been of high quality because of the difficulty in controlling the reactions of the highly reactive photolyzed species. By using a pulsed IR carbon dioxide laser, however, high quality gallium arsenide thin films have been grown in a process that raises the temperature of the reacting gas immediately above the substrate (Jones 1985).

It has been observed that SiF_6 molecules, inert to Si at 25°C, can be induced to react with the solid by multiple photon excitation with a pulsed carbon dioxide laser to form SiF_4 and SF_4 products (Chuang 1981). Three or four photons are involved to vibrationally excite SF_6 molecules via process (1).

At 90°K, SF_6 adsorbs on the surface of Si with 1 or 2 monolayers of coverage. When irradiated by carbon dioxide laser pulses in resonance with SF_6 vibrations, a substantial number of SiF molecules are formed (process (1)) and desorb into the gas phase where they can be detected with a mass spectrometer (Chuang 1982b).

Laser interaction with a solid surface generally results in localized heating that promotes surface reactions. A focused Ar laser in the UV range has been shown to excite and heat a spot on a Si wafer to its melting temperature and thus induce etching by chlorine or HCl gas at a rate approaching 10 µm/sec (Ehrlich et al. 1981).

Studies at MIT (Ehrlich and Tsao 1983a) have shown that the reaction rate of a microchemical process induced by a focused laser beam can significantly exceed those of conventional diffusion-limited reactions for an extended heat source, in which reaction rates at a gas-solid interface are generally limited by the gas phase diffusion of reactants and products, and by the reaction rates on the surface itself. The reactant flux channeled into the reaction zone increases as the dimension of the reacting area decreases to val-

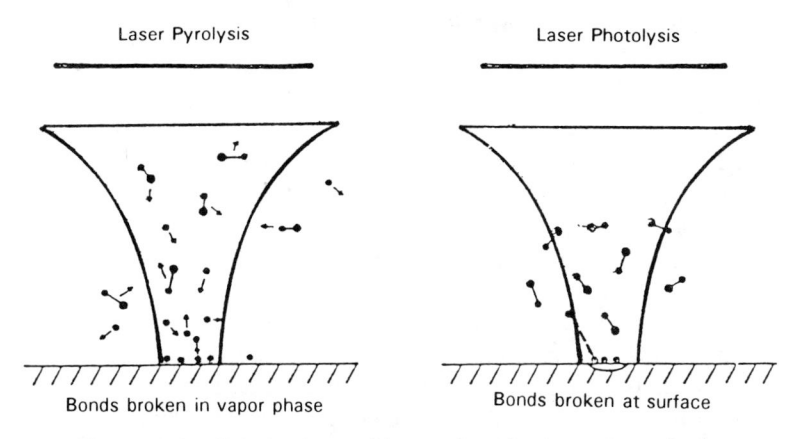

Figure 9.2. Comparison of laser photolysis and pyrolysis.

Laser Pyrolysis — Bonds broken in vapor phase
Laser Photolysis — Bonds broken at surface

ues small compared with the gas diffusion distance due to the enhancement of mass transport via three-dimensional gas phase diffusion. In comparison, the reaction flux on a planar surface is limited by one-dimensional gas phase diffusion.

Processes (1) and (2) are commonly referred to as laser pyrolysis and laser photolysis respectively, as illustrated in Figure 9.2 (Oprysko 1986). With *laser pyrolysis*, the chemical reaction occurs at the surface, while the volatile components remain in the vapor phase. Using this technique, features as small as 1 µm have been deposited. This method has a number of advantages:

- Depending of the optical delivery system, a compact 50 mW air cooled argon ion laser suffices as a source.
- Argon ion laser output is tunable and stable over long periods of time.
- Since the pyrolytic deposition utilizes a visible laser, no expensive uv or ir optics are required.

In *laser photolysis*, a laser is used to decompose an organometallic molecule such as $Cd(CH_3)_2$ in the vapor phase. Although studies have shown that submicron processing is possible, this method has several disadvantages:

- Second harmonics that are generated lead to inefficiencies in the process requiring a large, high-power argon laser.
- These second harmonics exhibit an output that fluctuates with time, requiring periodic adjustments.
- An argon laser operating at 257 nm requires costly, high-numerical aperature, long-working-distance quartz optics.

9.3. SEMICONDUCTOR PROCESSING

9.3.1. Laser Doping

A UV-laser doping technique has been developed by XMR Corp. (Santa Clara, Ca) in which a XeCl excimer laser is used to form controllable diffusion junctions in substrates 40 to 200 nm deep (Lineback, 1986). The laser forms a shallow molten layer on the substrate surface contained in a chamber filled with a dopant such as pyrolytic diborane (Skidmore, 1988). The laser dissociates the gas causing boron to diffuse into the molten silicon. The melting and recrystallization occurs in about 200 ns. By rastering the laser, the surface of a 150 mm wafer can be covered in 40 seconds.

Laser-assisted doping has been demonstrated at Stanford University where NMOS transistors with minimum feature sizes of 0.9 µm were fabricated. Work is in progress to fabricate 0.5-µm CMOS devices.

A variation of doping is used to create circuit links between metallized layers. Developed at MIT, a millisecond, 2-W argon ion laser melts an amorphous silicon layer separating two metal layers. A dopant is allowed to diffuse into the melted silicon, creating a 2-µm thick continuity path of 100 ohms.

9.3.2. Wafer Inspection

Wafer inspection is performed to determine process-generated defects, registration, and critical dimensions. Companies offering laser-based illumination sources for image processing are Lasertec and Nikon.

A new approach to defect detection is based on holographic technology and developed by Insystems. The system, a stopgap mask inspection tool that is marketed primarily for a wafer inspection, is based on optical processing technology rather than the digital processing of scanned data employed by other systems described above. As a result, the system can inspect more quickly (minutes as opposed to hours) with a sensitivity to 0.5 µm. With this method, a hologram is generated by illuminating the pattern with a laser that is then compared to the hologram of a perfect pattern. The entire wafer, both surface and subsurface, is examined in under 30 minutes compared to 30 hours by conventional techniques. The resolution is 0.2 µm.

A 1-W argon-ion laser operating at 514.5 nm is used to form the holographic image in a three step process:

- A reference filter plate representing the periodic structures of the wafer circuit pattern is first made.
- The wafer is then exposed to the laser and imaged through the filter to form a holographic image of the wafer.

LASER PROCESSING

- The holographic image is reconstructed and displayed on a CCD video camera.

Another approach to inspection, improvements over optical and scanning electron microscopes (SEM) is confocal laser scanning illustrated in Figure 9.3. A point source of a short-wavelength laser is focused onto the wafer through a high NA objective lens. The light reflected from the wafer is refocused by the same objective lens and separated by a partially reflected mirror. The reflected component is focused on a pinhole aperature and detected by a photomultiplier tube. By rastering the beam across the wafer, reflected light intensity from each point is analyzed by a computer and displayed from a digital frame store in a video screen (Keeler 1991).

This type of system is available from SiScan and Nikon, which use an Ar ion laser at 488 nm to match resist dye absorption for maximum contrast. Bio-Rad and Heidelberg Instruments use a 5-mW HeNe laser to achieve a resolution of 0.3 µm.

White-light confocal systems can use either a light source with a filter for specific wavelengths or the whole broadband of the light source for white light. These systems, less expensive and computer intensive than monochromatic laser systems, also experience a much smaller degree of interference (Bond 1992). Available from Carl Zeiss, white-light confocal systems are also better for measuring resist and other features.

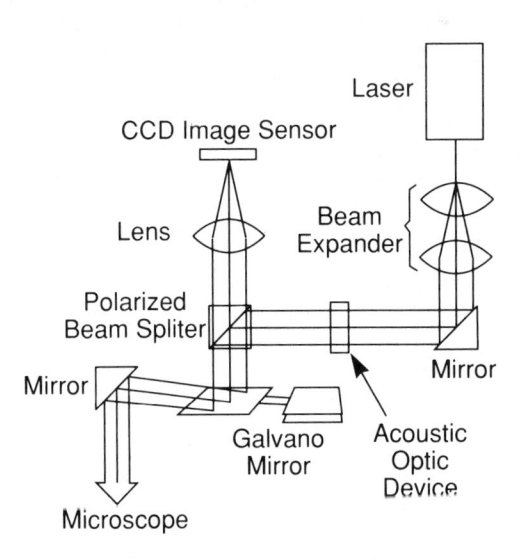

Figure 9.3. **Schematic of confocal scanning laser microscope.**

9.3.3. Wafer Repair

Circuit damage is a problem when prototypes are fabricated with faults in the aluminum interconnection lines. Work at Sandia National Laboratories has resulted in a method of repairing these defects with a KrF laser. For cutting, the laser is aimed at the defect through a quartz window of a chamber filled with chlorine gas. Between 20 and 30 pulses are made, each of 0.1 Joules of energy/cm^2. The laser cracks the aluminum oxide coating, exposing pure aluminum that reacts with the chlorine and is pumped away. For deposition, a CW argon ion laser of 0.4 W is focused on the wafer located in the chamber containing diborane and silane gases. Localized heating of 700–800°C causes the gases to decompose and deposit conducting polysilicon. Work at MIT's Lincoln Laboratory focuses on the deposition of tungsten.

One of the earliest applications of lasers was for thick and thin film resistor trimming of hybrid circuits. With the development of 64-Kbit DRAMs, yield improvement was enhanced with the incorporation of redundant memory. This link-cutting method is used for programming redundant memory cells that replace bad bit locations (Richardson and Swenson 1989).

XLR's (Norwell, MA) Model 525 is based on a Q-switched, diode-pumped solid state laser manufactured by Spectra-Physics (Mt. View, CA) has a link disconnect rate in excess of 2000 links/sec with a guaranteed beam positioning accuracy of ±0.45 µm. ESI's (Portland, OR) Model 9000, a diode-pumped Q-switched Nd-YLF laser also manufactured by Spectra-Physics, features a Unix-based workstation, robotic wafer handling for up to 200 mm wafers, and a self-contained cleanroom environment. It is geared for submicron devices, offering a beam positioning accuracy of ±0.5 µm. Teradyne Laser Systems' (Boston, MA) Model M218 uses a 50-µJ Nd:YLF manufactured by Lightwave Electronics (Mt. View, CA). The system has 7-ns pulses and can process up to 2,000 devices per hour (Forrest 1989).

Yield on 256 KBit devices has improved from 35% to 65% using this approach. The $700,000 repair systems can have a payback in as little as three days. According to ESI, each salvaged die represents a net profit of $6–$10, and one machine can repair 20,000–30,000 dies per day. The greatest problem is the removal of debris generated by the laser, which can contaminate the device.

XRL, ESI, and Teradyne Laser Systems offer versions of their systems for thin-film resistor trimming. Device performance is monitored during laser trimming as material ablated from the device adjusts the resistance of specific circuits. ESI's Model 6000 uses a diode-pumped Q-switched Nd-YLF laser manufactured by Lightwave Electronics.

Although FIB has been the exclusive domain of mask repair, it is now entering into the circuit repair arena. FIB suppliers have recognized the lim-

LASER PROCESSING

ited potential of this market and have introduced modified versions of their mask repair systems for use in IC modification. Micrion, which markets a system for mask repair, has the Micrion DMOD, an automatic system to modify ICs directly on the wafer.

For use in IC modification, FIB can remove and deposit material on a circuit in a similar manner to mask repair; minor changes in a chip design could be made and tested, instead of changing masks and running new silicon. FIB can microsection ICs by milling away areas of the layered structure with the high energy ions. The area of interest is observed on the CRT of the FIB system by SIM (Secondary Ion Microscopy). Material contrast is much greater with SIM than with SEM, and there is no special sample preparation needed.

9.3.4. Circuit Failure Analysis

Lasers are also being used to cut metal conductor traces for the analysis of circuit designs and defects (Forrest 1989). The technology, initiated by Florod (Gardena, CA) using a xenon laser connected via a fixed optical transfer arm, has now been incorporated by every wafer probe-station manufacturer. Alessi (Irvine, CA) followed with a Nd:YAG laser cutter mounted directly on the microscope of the probe station. Other companies offering this technology include Hoya Optics (Fremont, CA) with a YAG-based system, Micromanipulator (Carson City, NV) with a YAG-based system, Signatone (Gilroy, CA) with a Florod xenon laser-based system, Wentworth Labs (Brookfield, CT) with a Florod xenon laser-based system, and Karl Suss with a Pacific Laser xenon laser-based system.

Current technology in laser-based failure analysis involves removing layers of the circuit to determine the nature of the failure. The newest technology involves the use of the laser to analyze passivation-coated circuits. Future systems will incorporate the laser-based probe stations with CAD information. Knights Technology (Cupertino, CA) has introduced a computer-aided verification package called Excalibar that is in use by Alessi, Signatone, and Wentworth Labs. Electron-beam probes will offer laser-based systems stiff competition in the very near future.

Reliability and failure analysis of a thin film are also determined with the aid of a FIB. The microstructure and grain size distributions that affect the electromigration of an aluminum film can be monitored and analyzed using SIM. Previously only transmission electron microscopy (TEM) had the capability of observing such microstructures, requiring a great deal of time in sample preparation. It now appears that SIM contrast exceeds TEM contrast.

9.3.5. Particle Detection

Particulate monitoring of air purity, airflow velocity, air pressure, harmful gases, microorganisms, humidity, and temperature in the cleanroom is required in order to determine the cause of mask and wafer defects during standard processing conditions. Continuous, real-time monitoring at each processing step enables accurate determination of the cause of wafer damage when it occurs. This type of monitoring equipment would include:

- Instantaneous particle counts of air adjacent to wafer

- Compact air sample intakes and detection mechanism so as not to interfere with processing

- Simultaneous recording of times and particle counts

- High particle detection accuracy

The optical particle counter is the most widely used method in cleanrooms. A light source illuminates a small viewing volume and a photodetector measures the scattered light from the individual particles as they pass through this illuminated volume. Lightscattering systems include monochromatic laser and polychromatic white light monitors illustrated in Figure 9.4 (West and West 1986).

Both systems operate nearly identically; the difference between the two is that the laser system is able to concentrate more energy in a smaller spot.

A large number of improvements have taken place with both laser and white light systems over the past few years, lowering the limits of sensitivity, and increasing repeatability. In principle, sensitivity limits vary from system to system, with the counting efficiency dropping sharply from 100% to 10–30% when a particle size of 0.35 μm is reached (Liu *et al.* 1987).

Sensitivity limits, which also depend on the gas flow rates, are:

- Laser-based systems

    — 0.1 μm for flow rates to 1 cfm
    — 0.3 μm for high flow rates, with an efficiency drop off at 0.4 - 0.5 μm for some particle counters.

- White light-based systems

    — 0.3 μm for low flow rates, with an efficiency drop off at 0.4 - 0.5 μm for some particle counters.
    — 0.5 μm for high flow rates, with an efficiency drop off at 0.6 - 0.7 μm for some particle counters.

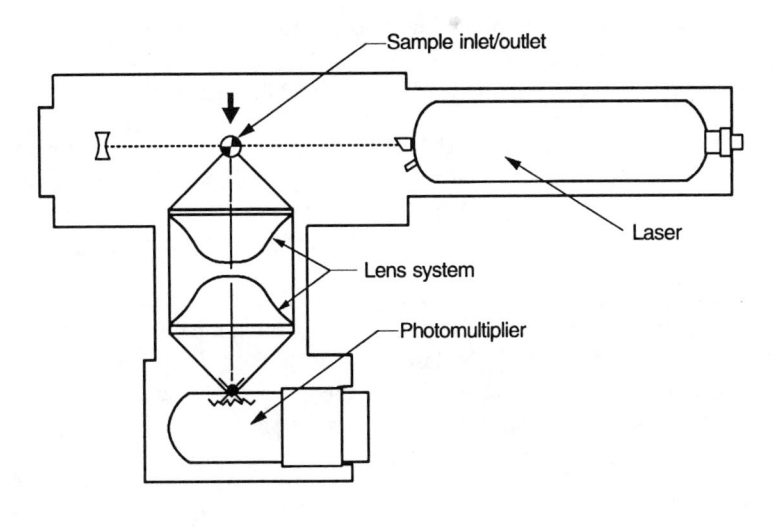

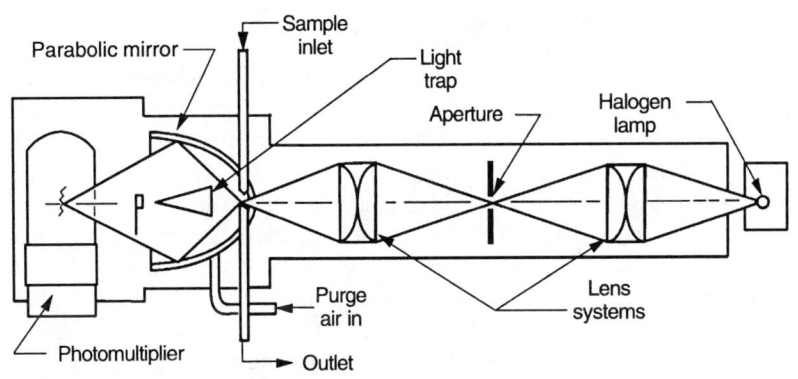

Figure 9.4. **Laser and white light-based particle measurement systems.**

Difficulties arise when flow rates are reduced in order to increase sensitivity causing:

- High signal-to-noise ratio, equivalent to 1 to 10 particles per cubic foot, rendering counts in a Class 10 cleanroom meaningless.

New advances in electronics has resulted in counters with reduced signal-to-noise levels, so that the counting efficiency may be even higher at lower flow rates.

Existing counters in numerous cleanrooms, however, do not have sophisticated electronics, so that the counting efficiency is affected by the age of the counter, and whether any dirt has entered the system.

- At low flows, sampling must be made over longer time periods, up to 1.5 hours. This represents a problem with personnel activity, which is a major source of contamination within the cleanroom. Measurements, show that contamination levels are directly related to personnel activity.

The monitoring of particulates on the surface of a mask or wafer employs a surface scan detection systems after the contaminant has condensed on the surface. A focused He-Ne laser beam scans the surface in a raster scanning pattern. On a contaminated surface, the laser beam is scattered and collected by a photomultiplier tube with a particle-size sensitivity of one μm.

9.3.6. Endpoint Detection

Detectors are generally integrated into the dry processing system to detect the end of the process. End-point detectors can be incorporated as add-ons, however, and the user should make a choice, based on the materials to be etched.

There are two widely used methods for endpoint detection as illustrated in Figure 9.5. Optical Emission offers the greatest flexibility and can either detect the average endpoint in a batch reactor or the endpoint of a single wafer. Any number of emission lines can be monitored to determine the endpoint. Several problems occur in etching silicon dioxide in which the CO emission line is often monitored. These lines (1) are fairly weak and the silicon dioxide on the wafer is small so that accurate monitoring is difficult, and (2) can be masked by other lines due to resist erosion. Optical emission also has the advantage that it can be used to yield other information besides endpoint, such as a diagnostic tool to observe contaminants or residue (Marcoux and Foo 1981).

Another tool is Laser Interferometry (Buste *et al.* 1979) in which a He-Ne laser beam is reflected off the wafer surface and detected by a photocell. There are two problems with this technique: (1) it only monitors one spot on a single wafer, and (2) the wafer must be aligned so that an open area is monitored. In some cases, this necessitates the need for a test area to be added to the wafer if the exposed area is too small.

Two other methods have proven successful and require no additional equipment: (1) change in pressure at constant flow rate of the etch gas (Fok 1980), and (2) change in the dc self-bias voltage at constant rf power (Brown *et al.* 1978).

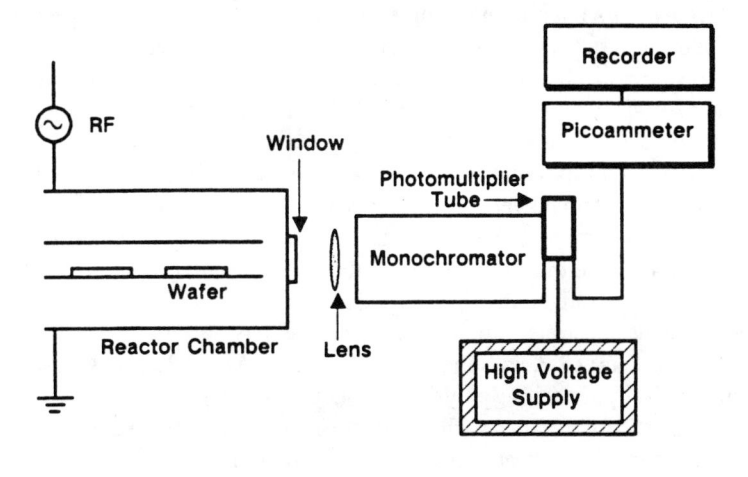

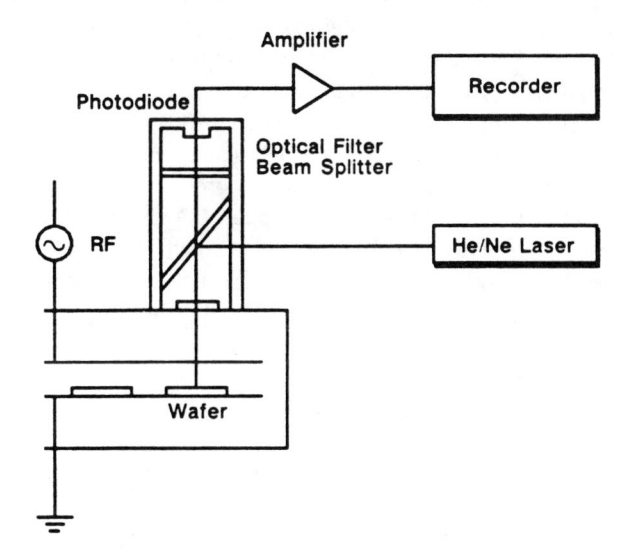

Figure 9.5. Methods of endpoint detection.

9.3.7. Laser-induced Deposition

Numerous studies have been made for the selective deposition of various materials for IC applications. A large increase in the steady-state rate relative to their conventional counterparts has been measured (Ehrlich and Tsao 1983b). As an example, pyrolytic silicon chemical vapor deposition (CVD) in a standard furnace has a maximum useful rate of <10^{-2} μm/sec but produces excellent electronic-grade polysilicon at rates >10^2 μm/sec in laser direct writing.

Tungsten on silicon has been selectively deposited by the reduction of WF_6 (Liu *et al.* 1985). In conventional CVD, the reaction occurs on the silicon surface at a temperature of 350°C. Using an argon laser, the reaction takes place between milliseconds to several seconds, depending of the scanning speed, spot size, and power of the laser beam.

At the University of Massachusetts, researchers have grown high quality gallium arsenide thin films with a carbon dioxide laser by adjusting the AsH_3 partial pressure, total pressure, and gas flow (Jones 1985). By exciting PH_3 above the substrate, superlattices of GaAsP have been grown.

Researchers at Sandia National Laboratories have used a CW Ar laser to repair circuits by depositing polysilicon conductors (Waller 1986). The wafer is placed in a chamber filled with a mixture of diborane and silane gases. The 0.4 W laser traces and heats a path over the insulating layer. When the path reaches 700 - 800°C, the two gases decompose and conducting polysilicon deposits on the path. At MIT Lincoln Labs, this process has been used to deposit thin films of tungsten on laser-written polysilicon interconnects and GaAs MMICs. WF_6 gas is exposed to the laser-written lines in a furnace at 400°C for five to six minutes to deposit films from 100 to 150 nm thick. Films are written at a rate of 2.5 mm/sec (Rose 1986).

The major commercial application of this process is the maskless fuse linking of application-specific integrated circuits (ASICs). Large semicustom circuits can be routed by the deposition of conductive boron-doped polysilicon links on chips using Ar lasers (Ehrlich *et al.* 1984). An array-programming system automatically calculates routing paths. Currently, 10 links per second can be formed.

9.3.8. Laser-induced Etching

There are two distinct processes in laser etching, material vaporization and laser-assisted chemical etching. Material vaporization is presently used for the custom tailoring of ASICs. LaserPath (San Jose, CA) developed a process for fabricating CMOS gate arrays using a YAG laser that can cut interconnections directly on arrays with several hundred thousand to a million laser flashes per die. The die is already packaged in the uncovered ceramic package with nitrogen flowing across the wafer. The laser technique has resulted in an elimination in the number of process steps compared to the conventional production cycle.

Laser-assisted chemical etching is accomplished in either the wet or dry mode. Researchers at MIT Lincoln Labs have demonstrated a wet etching process for Al that involves localized laser heating (< 200°C) of a surface

LASER PROCESSING

bathed in a capillary liquid etchant composed of 0.15 percent potassium dichromate in a 10 percent H_3PO_4/90% HNO_3 solution that is trapped between the Si wafer and a thin cover glass (Ehrlich *et al.* 1984). At Sandia National Laboratories, laser-assisted dry etching is used to cut aluminum lines. The wafer is contained in a vacuum chamber filled with chlorine gas. A KrF excimer laser is aimed through a quartz window and pulsed between 20 and 30 times per second onto the aluminum line. This heats the aluminum, cracking the 2-nm oxide film so that chlorine and aluminum could spontaneously react to form volatile aluminum chloride.

9.3.9. Laser Planarization

VLSI devices with submicron feature sizes are often fabricated from up to 15 mask steps. Each of the mask steps that require the deposition of a metal generates a mesa structure which, when subsequently coated with a dielectric material, results in the surface with the appearance of hills and valleys. This presents problems with depth of field during subsequent lithographic processes on uneven surfaces. Also, processing steps also give rise to hillocks and surface roughness.

Laser planarization of the silicon surface has been demonstrated by XMR using a XeCl excimer laser to remelt a 1-µm-thick aluminum film (over 1500 A of TiW) deposited over oxide steps and via patterns with geometries ranging from 4-µm pitches to 0.75-µm via holes. The energy of the beam is 2-6 joules/cm^2 with a ±3% uniformity over more than 90% of the beam diameter. In less than 50 ns, the Al film is planarized with a mean surface profile of ±40 nm. The 0.75-µm wide, 1.5-µm deep vias were filled completely as aluminum flowed into the vias by multiple pulses of the laser, achieving uniform planarization (Forrest 1990).

In addition to planarizing the surface, the resultant heating of the aluminum increases size of the grain boundaries of the films from 0.2 µm to 5–10 µm, making them less susceptible to defect formation. The planarization process apparently reduces the stress that causes hillock formation.

There are numerous methods used for planarizing an oxide surface before the film is deposited:

- Thermal reflow
- Etchback
- Polyimides
- TEOS (tetraethyl orthosilicate)
- Sacrificial layer etch back

- Bias sputtered dielectrics
- PECVD of dielectrics
- Spin-on-glass
- Lift-off
- Chemical-mechanical polish (CMP)

There are alternatives to planarization of a metal surface:

- CVD tungsten
- Bias-sputtered aluminum
- Selective tungsten deposition
- Stud formation
- Dielectric implantation

Contact via filling can have a large effect on the industry, reducing the need for alternative CVD refractory thin film deposition systems that are currently used for via filling. The technology must compete with selective deposition of CVD tungsten, whereby the W film selectively deposits on silicon but not on oxide. The process is slow, however, and nucleation sites on the oxide can form whereby W is deposited there as well.

9.3.10. Laser Pantography

The above reactions, lithography, deposition, and etching have been used for the fabrication of complete devices in a process called laser pantography. The lithography step is generally a direct-write process that avoids the use of photoresist. There are two methods for the lithography step:

- Direct laser writing in which a focused beam is used to induce direct deposition and etching at dimensions as small as 0.2 μm (McWilliams *et al.* 1983). This is shown in Figure 9.6.
- Excimer laser projection, similar to a conventional reduction stepper, in which the laser beam is passed through a reticle and projected onto the wafer housed in a vapor cell (Ehrlich 1985). This method is shown in Figure 9.7.

9.3.11. Packaging

After a wafer is processed, it is cut into individual die to be subsequently packaged. Laser IC separation was first used in the early 1970s as a method to

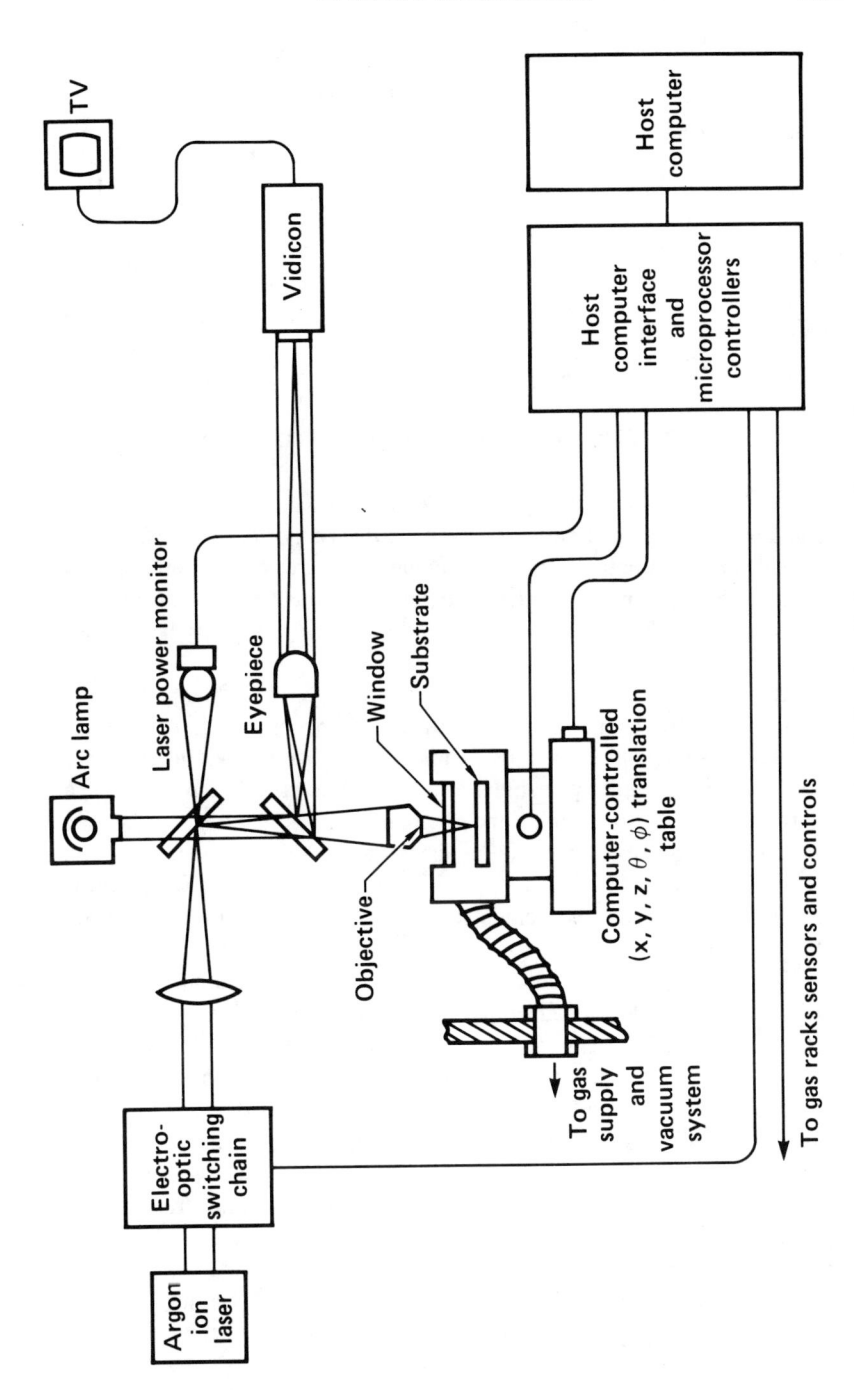

Figure 9.6. Schematic of direct write laser pantography.

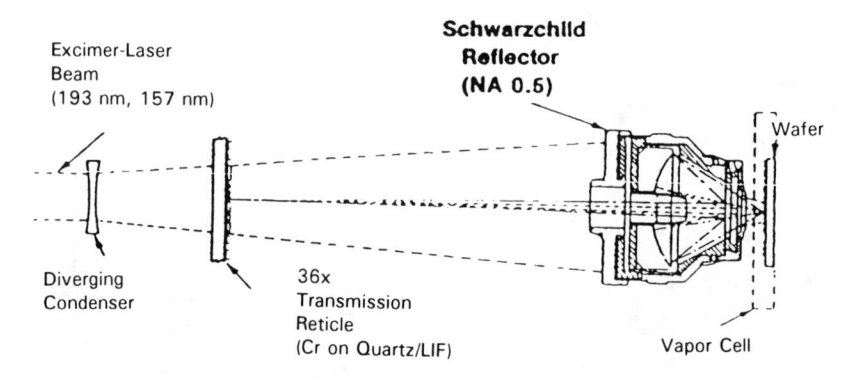

Figure 9.7. Schematic of laser projection microchemistry.

eliminate diamond saw dicing. However, residues deposited on the front and back side of the wafer result in circuit damage and wire bonding failures. The advent of the high-speed, rotating blade diamond saw that eliminated these problems has resulted in a decreased use of laser dicing for chips that are neither rectangular or square.

Researchers at General Electric received a two-year, $790,000 Strategic Defense Initiative contract to use direct laser deposition to connect unpackaged semiconductor chips mounted directly on polyimide or other PCBs. Using either epoxy or solder as an adhesive, the board is placed in a gas-tight chamber filled with a metal-containing gas. An excimer laser is focused on the substrate, dissociating the gas and resulting in the deposition traces of tungsten, aluminum, copper, molybdenum, zinc, and gold as narrow as 1 µm. The advantages include maskless operation and deposition on non-flat wafers.

Researchers at Microelectronics and Computer Technology Corp. (Austin, TX) have developed a tape automated bonding (TAB) method that utilizes a pulsed Nd:YAG laser to bond 2-mil leads to I/O pads at a rate of 45 lead bonds/sec and a rate of 65 lead bonds/sec in the near future. The TAB tape is a 35 mm polyimide film bonded to copper. The copper is etched to form individual leads and the tape is positioned over the IC's bonding pads. A pulsed YAG laser is the most suitable in terms of material absorption at 1.06 µm and the ability to shape the laser pulse to maximize bond strength. The company licensed the technology to ESI for a complete commercial system that became available in 1990 (Swenson 1990).

In another bonding process, the Vanzetti Systems' ILS series of intelligent laser soldering systems can produce perfect joints every time by adjusting the heat they deliver to individual joints (Rieley 1990). As shown in

LASER PROCESSING

Figure 9.8, an IR detector monitors the formation of each joint as the Nd:YAG laser beam heats the solder to its melting point. This is an improvement over traditional laser soldering systems that apply the same amount of heat to each joint regardless of variations in the amount of solder at each joint or differences in surface dimensions. Furthermore, inspection takes place simultaneously with joint formation, as the "thermal signature" (the inspection certificate) appears at the detector output the very instant when the joint is completed. Laser soldering is best suited to large, fine-pitch, high-lead-count packages.

In addition, the laser solder process is an improvement over conventional IR and vapor-phase reflow methods that often result in undesirable large-grain recrystallization because of the length of time required to cool the joint to solidification, and because boards are held above the solder's melting point for a relatively longer period, which can affect device reliability.

The fail-safe feature of the ILS eliminates the risk of burning the components or the board from accidental misalignment or surface contamination.

Intel is developing a laser-based metal trimming system to fabricate lead frames with Jade Corp. These lead frames are used to connected the diced chips through the plastic encapsulant to the printed circuit board.

9.3.12. Laser Marking

Laser marking is used as an inventory control during processing. A problem that exists on the fab line is that every wafer looks exactly alike. Mistaken identity can result in misprocessing. Laser marking involves writing a multi-digit number on each wafer as it starts into the manufacturing line. The mark, placed in the surface layer by a medium-powered laser, is indelible so that it survives all the processes the wafer undergoes during fabrication.

Lasers can also be used to mark packaged wafers. Printed codes on the package tell the user product type, country of origin, and other essential information about the part that they are using. They are also used by the manufacturer for inventory control. The traditional method uses ink, deposited either by direct printing, offset printing, gravure printing, or ink jet printing. A typical system costs $5,000.

Laser marking, on the other hand, utilizes either a TEA (transversely excited atmospheric) carbon dioxide or Nd:YAG laser. The TEA laser is permanent and fast, and, unlike ink, can mark moving devices and cannot smear. Lumonics (Kanata, Ont., Canada) has a TEA carbon dioxide laser that will operate at 100 Hz, marking character by character at a rate of 100 characters/sec. However, it will not work with metal parts as will the Nd:YAG laser. Both systems are expensive, costing above $30,000, and are limited by low contrast and difficulty in demarking.

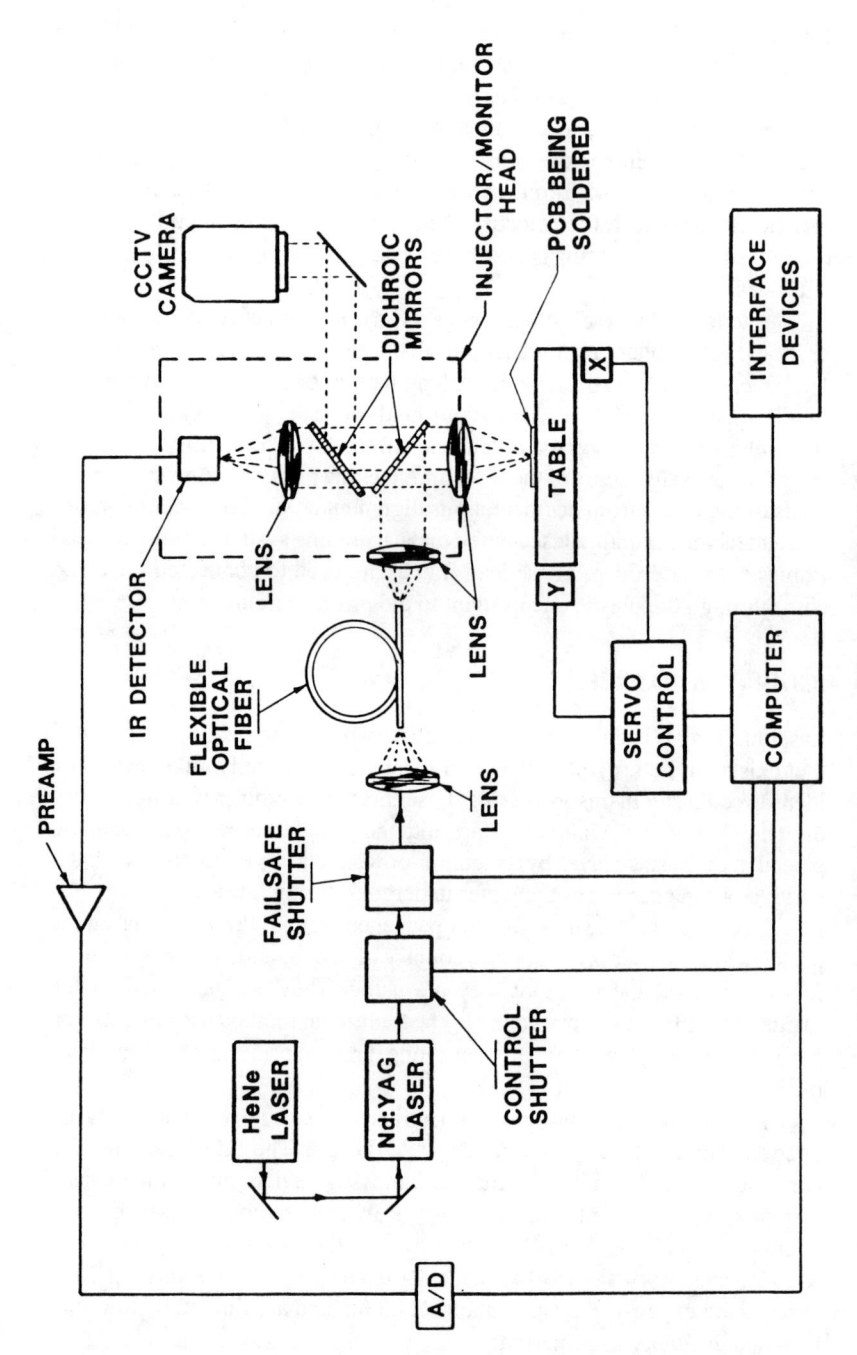

Figure 9.8. Schematic of intelligent laser soldering process.

LASER PROCESSING

During marking, the image of a stencil is projected onto the surface of the part to be marked. A single millisecond pulse is used to vaporize the surface. Only 3% of packages currently use this method. Nevertheless, it is expected that the advantages of laser marking, described above, will result in an increased usage of this method. An additional feature of laser marking under evaluation today is the capability of integrating testing of the circuit with marking. This process has the added benefits of:

- A single system with a small footprint
- Reduction in the risks of electrostatic discharge due to handling
- Reduction in operator error resulting in mishandling and improper inventory control

Marking with a 1.06 μm Nd:YAG laser can cause sever damage to the silicon substrate, as silicon is partially transparent to IR radiation. Local damage could spread throughout the substrate causing dislocations. Baasel Lasertech (Starnberg, Germany) uses a frequency-doubling approach at 532 nm, a wavelength at which silicon is opaque and damage is confined to the surface.

Laser marking systems also include Adcotech's Laser series, Automated Industrial Systems' AS4002RR, Control Laser's Instamark, Laser Identification Systems' Wafermark series, Lumonics' LaserMark and WaferMark series, Photon Sources VM series, and U.S. Laser's 450X series.

9.3.13. ASIC Processing

The gate array category is being encroached on the low-end by PLDs with densities above 10,000 gates and faster turnaround times available with standard cells. Laser-based systems could enable gate-array vendors to compete with these technologies by reducing the turnaround time for a typical gate array from weeks to hours—the specific time being dependent on the numerous types of laser-based systems available.

There are two fundamental methods of fabricating these quick-turnaround devices: subtractive, where the laser cuts interconnects on the metallization levels for customizing; and additive, whereby the metallization is deposited by the laser. Several companies have developed equipment for this fabrication described below (Burggraaf 1988) (Markowitz 1990).

Laserpath

Laserpath (Sunnyvale, CA) was one of the initial developers of this concept. The company stopped making the system in mid-1988 due to high costs developing of the machine. A prototype of a gate array could be completed in one day, compared to 6 to 10 weeks for a traditional approach. Using a subtractive method, the company used a Model 800B Nd:YAG laser system (ESI) for its line cutting operation.

Laserpath used pre-processed and pre-packaged generic arrays of up to several thousand gates that were interconnected. The lines were then cut by the laser to form the required logic functions. Initial Laserpath products ranged from 880 to 3,600 gates.

Lasarray

Lasarray (Biel, Switzerland and Irvine, CA) uses a Liconix HeCd laser generator to customize a wafer through a positive resist exposure process similar to direct-write electron beam lithography. After writing, the full wafer pattern from a CAD program is downloaded to the system, which is also capable of a subtractive metal etch step.

The only modification to wafers is a grid mask that functions as a laser guide. The typical size of the holes in the grid is 1 µm. The laser exposes metal lines in the resist, connecting holes and exposing metal lines in the resist, thus linking gates in a desired manner.

The system will support metal pitches under 6 µm and can handle 16 different designs on a wafer for a cost of $12,000 to $15,000 plus an non-recoverable engineering (NRE) cost of $30,000.

Lasarray offers the system for sale consisting of three transportable containers measuring 7.5 × 9 × 3.6 meters:

- Container 1 contains a Class 10 area that houses the laser pattern generator, a proximity mask aligner for the metal personalization, a plasma deposition (passivation) and etching (passivation and metal) system, a resist processor, and a cleaning and drying module.

- Container 2 contains a Class 10,000 for wafer dicing, packaging and testing.

- Container 3 houses air conditioning, electrical equipment, and gas handling.

The complete system is priced at $4.2 million, and a second direct-write laser unit is available for $600,000.

Lasa Industries

Lasa Industries (Santa Clara, CA) uses an additive method for personalization of the gate array. The fabrication process begins with an uncommitted array processed by conventional methods through aluminum metallization. The wafer is diced and mounted in unopened special packages and shipped to the user's facility where it is placed in a self-contained unit measuring 76 × 84 × 33 inches. This unit contains the argon laser, computer controls, hardware, and chemical and gas handling equipment for processing the chip.

Using reformatted GDS-II output, a robotic system in the unit moves the devices to a chamber for the direct-write deposition of the first layer of tungsten metal interconnection, deposition of a passivation layer, direct etching of vias in the layer, and direct-write deposition of a second layer of tungsten for final interconnection. A robot then transfers the device to the final assembly area where ceramic lids seal the packages.

The QT-GA system is priced at $3 million and can process a 1,000-gate array with two levels of metallization within hours. Three-level metal capabilities are available. The system is capable of processing 1-μm geometries and arrays up to 100,000 gates are feasible.

Chip Express

Chip Express (Santa Clara, CA) a spinoff of Elron Industries (Haifa, Israel) has developed the Quick gate array that uses a subtractive method of customizing the circuit as does Laserpath. The company uses a Nd:YAG frequency-doubled laser controlled by a host computer based on a Motorola 68020 microprocessor housed in a compartment measuring 6 × 4 × 5 feet.

Unlike other approaches that use standard or slightly modified blank arrays, the Elron process requires base arrays that are modified in four areas: input/output cells, core logic array, core routing, and core interface with the I/O cells. A special calibration region on the die must also be included to calibrate the laser.

The system, priced at $500,000, is capable of prototyping a 2,500-gate double metal CMOS array in 40 minutes and a 10,000 gate array in 90 minutes. To insure high yields, the fabricated arrays won't have as many usable gates as arrays corresponding to mask-programmable systems. Specially modified arrays, based on Seiko's 2-μm, 6000-gate family, have 5,500 to 5,900 gates available after processing on the Quick system.

Elron departed from Laserpath's approach of cutting metal and now uses the subtractive approach to remove photoresist above a metal. The metal is then etched in liquid etchants or by plasma etching.

REFERENCES

Bond, J., 1992: "Linewidth measurements struggle with submicron processes," *Test & Measurement World* **12** (5): 65–71.

Brau, C.A., 1984: in *Excimer Lasers*, C.K. Rhodes ed., 2nd edition (Springer-Verlag Berlin, New York).

Brown, H.L., G.B. Bungard, and K.C. Lin, 1978: "Applications of mass spectrometers to plasma process monitoring and control," *Solid State Technology* **21** (7): 35–38.

Buste, H.H., R.E. Lojos, and D.A. Kiewit, 1979: "Plasma etch monitoring with laser interferometry," *Solid State Technology* **22** (2): 61–67.

Burggraaf, P., 1988: "Laser-based pattern generation," *Semiconductor International*, **11** (5): 116–121.

Chuang, T.J., 1981: "Multiple photon excited SiF_6 interaction with silicon surfaces," *J. Chem. Phys.* **74** (2): 1453–1460.

Chuang, T.J., 1982a: "Laser-enhanced gas-surface chemistry: basic processes and applications," *J. Vac. Sci. Technol.* **21** (3): 798–806.

Chuang, T.J., 1982b: *Vibrations at Surfaces*, R. Caudeno, J.M. Gilles, and A.A. Lucas, eds., (Plenum Press, NY).

Ehrlich, D.J., 1985: "Early applications of laser direct patterning: direct writing and excimer projection," *Solid State Technology* **28** (12): 81–85.

Ehrlich, D.J., R.M. Osgood, Jr., and T.J. Deitsch, 1981: "Laser chemical technique for rapid writing of surface relief in silicon," *Appl. Phys. Lett.* **38** (12): 1018–1020.

Ehrlich, D.J. and J.Y. Tsao, 1983a: in *Laser Diagnostics and Photochemical Processing for Semiconductor Devices*, R.M. Osgood, S.R.J. Brueck, and H.R. Schlossberg, eds., (North-Holland, Elsevier, NY).

Ehrlich, D.J. and J.Y. Tsao, 1983b: "Laser direct writing for VLSI," in *VLSI Electronics: Microstructure Science,* Vol. 7, (Academic Press, NY).

Ehrlich, D.J., J.Y. Tsao, D.J. Silversmith, J.H.C. Sedlacek, R.W. Mountain, and W.G. Graber, 1984: "Laser micromachining techniques for reversible restructuring of gate-array prototype circuits," *IEEE Electron Dev. Lett.* **EDL-5** (2) 32–35.

Elza, D. 1985: "Lasers take to the factory floor," *Photonics Spectra* **19** (3): 75–78.

Fok, T.Y., 1980: "Plasma etching of aluminum films using CCl_4," Spring Meeting Electrochemical Society, St. Louis, MO.

Forrest, G.T., 1989: "Lasers move into semiconductor production in a big way," *Laser Focus World* **25** (9): 99–110.

Forrest, G.T., 1990: "Lasers are proving cost-effective in electronics production," *Laser Focus World* **26** (5): 161–170.

Jones, K.A., 1985: "Laser assisted MOCVD growth," *Solid State Technology* **28** (10): 151–156.

Keeler, R., 1991: "Confocal microscopes," *R&D Magazine*, Vol. 33, No. 5, pp. 40–42.

Lineback, J.R., 1986: "How lasers will give chip making a big boost," *Electronics* **59** (1): 70–72.

Liu, B.Y.H. and W.W. Szymanski, 1987: "Counting efficiency, lower detection limit and noise level of optical particle counters," Proc. Inst. of Environmental Sciences, pp. 417–421.

Liu, Y.S., C.P. Yakymyshyn, H.R. Phillip, H.S. Cole, and L.M. Levinson, 1985: "Laser-induced selective deposition of micron-size structures on silicon, " *J. Vac. Sci. Technol.* **83** (5): 1441–1444.

Marcoux, P.J. and P.D. Foo, 1981: "Methods of end-point detection for plasma etching," *Solid State Technology* **24** (4): 115–122.

Markowitz, M.C., 1990: "Fast-turnaround ASICs," EDN, Sept. 3, pp. 124–131.

McWilliams, B.M., I.P. Herman, F. Mitlitsky, R.A. Hyde, and L.L. Wood, 1983: "Wafer-scale laser pantography: fabrication of n-metal-oxide-semiconductor transistors and small-scale integrated circuits by direct-write laser-induced pyrolytic reactions," *Appl. Phys. Lett.* **43** (10): 946–948.

Oprysko, M.M., M.W. Beranek, D.E. Ewbank, and A.C. Titus, 1986: "Repair of clear photomask defects by laser-pyrolytic deposition," *Semiconductor International* **9** (1): 90–100.

Plummer, H., 1985: "The excimer laser: 10 years of fast growth," *Photonics Spectra* **19** (5): 73–82.

Richardson, T. and E. Swenson, 1989: "Laser redundancy boosts memory chip output," *Photonics Spectra* **24** (6): 133–136.

Rieley, D., 1990: "Making solder joints by laser," *Electronic Packaging & Production* **30** (1):. S48–S51.

Rose, C.D., 1986: "Laser writing takes a step forward," *Electronics* **59** (23): 15.

Skidmore, K. 1988: "Excimer lasers user for shallow junction formation," *Semiconductor International* **11** (3): 30.

Swenson, E.J., 1990: "Lasers expand reach in microelectronics production," *Photonics Spectra* **25** (11): 157–160.

Waller, L., 1986: "Cut-and-patch lasers speed chip repairs," *Electronics* **59** (24): 19–20.

West, G.C. and J. West, 1986: "Airborne contamination monitoring in the microelectronics industry," *Microelectronic Manufacturing and Testing* **9** (5).

INDEX

Acids 25, 30
Application Specific Integrated Circuits (ASICs) 21, 22, 49, 51, 76, 86, 97, 99, 120, 136, 216, 223
Automation 4, 15, 19, 147, 148, 150, 151, 166, 181, 198

Chemical delivery/dispensing 4, 5, 40, 45, 46
Chemical distillation 31, 47, 49, 51, 52, 56, 68
Chemical reprocessors 27, 46–51, 56
Chemicals (gases)
  carrier/bulk 61, 71
    argon 61, 65, 67, 68, 71, 162, 167, 175, 176
    helium 162
    hydrogen 61, 65, 67, 68, 71, 165, 191
    nitrogen 61, 65, 67, 68, 71, 162, 167
    oxygen 61, 65, 67, 68, 71, 165
  specialty 61
    ammonia 64, 66, 69, 71, 109
    arsine 64–67, 71
    diborane 62, 71, 109, 208, 210
    phosphine 61, 62, 64, 66, 67, 71
    silane 61, 62, 64, 66, 67, 69, 71, 210
Chemicals (liquid)
  acetic acid 26, 38, 41, 42, 52
  ammonium fluoride 26, 38, 41, 42, 44, 52

ammonium hydroxide 26, 38, 41, 42
buffered oxide etchants (BOE) 26, 41, 42, 52
hydrochloric acid 26, 38, 41, 42, 169
hydrofluoric acid 26, 38, 41–44, 51, 52, 120, 142, 169
hydrogen peroxide 26, 38, 41, 42, 47, 50
isopropyl alcohol (2-propanol) 38, 40
nitric acid 26, 38, 41–43, 52, 217
phosphoric acid 26, 38, 41, 42, 52, 217
sulfuric acid 26, 38, 42, 43, 47, 50–52
Circuit failure analysis 211
  scanning ion microscope (SIM) 211
  transmission electron microscope (TEM) 211
Cluster tools 7, 16, 181

Defects 3, 10, 13, 22
De–ionized water 3, 39, 53
Deposition gases/liquids
  arsine (AsH_3) 216
  carbon tetrachloride (CCl_4) 184
  molybdenum hexafluoride (MoF_6) 190
  phosphine (PH_3) 72, 216
  silane (SiH_4) 72, 178, 187, 188, 195
  tri–isobutyl aluminum (TIBA) 186
  tungsten hexafluoride (WF_6) 188, 190, 191, 216

Deposition methods 173
  chemical vapor deposition (CVD) 4, 67, 174, 178, 181, 185, 188, 189, 193, 196, 216
    atmospheric pressure (APCVD) 174, 178, 179, 193
    laser activated (LACVD) 174, 181, 196, 215, 216, 218, 220
    low pressure (LPCVD) 174, 178, 180–182, 186, 188, 190, 193
    plasma–enhanced (PECVD) 174, 181, 183, 184, 188, 190, 193, 194, 196, 197, 218
    sub–atmospheric (SACVD) 184
  electron cyclotron resonance (ECR) 7, 193
  evaporation 173, 174, 185, 196, 197
    electron beam 173, 175, 188
    ionized cluster 173, 174, 176
    resistance 173, 174
  sputtering 173, 175, 185–187, 189, 196, 197, 218
    dc/rf diode 173, 175, 176
    dc/rf magnetron 173, 175–177, 186
    ion beam 173, 176, 177, 184
Deposition reactors
  batch systems 196–198
  multi–chamber systems 197, 198
  single chamber systems 197
  single wafer systems 196, 197
Dielectric films
  BPSG 7, 178, 192–194
  BSG 178, 192
  nitrided TiW 192
  polyimide 161, 164, 217
  PSG 193, 195
  silicon dioxide (SiO_2) 117, 161, 164, 165, 181, 190–192, 194, 214
  silicon nitride (SiN) 117, 161, 164, 194, 195
  silicon oxynitride 181, 194, 195
  titanium nitride (TiN) 7, 187, 192
  TEOS/ozone 7, 193–195, 217
Diffusion 91

Dry processing 25, 27, 28, 152, 161, 166, 186
DUV/UV 73–76, 81, 85–87, 90, 91, 103, 114, 132, 138, 141

E–beam resists 100, 140
  COP 140
  PBS 140
  PMMA 30, 113
  single layer 100
Endpoint detection 147–153, 166
  laser interferometry 214
  optical emission 214
Etch profiles
  anisotropic 152, 153, 162–164, 166
  isotropic 153, 162
Etchants (gas)
  boron trichloride (BCl_3) 162, 164, 169
  carbon tetrachloride (CCl_4) 162, 164, 165, 168
  CCl_2F_2 167
  C_2F_6 163
  CHF_3 165
  chlorine (Cl_2) 149, 161–164, 167, 169, 206, 210, 217
  fluorine (F_2) 161–163
  fluorocarbons 162, 165
  HF 163, 165
  HCl 72, 206
  NF_3 163
  silicon tetrachloride ($SiCl_4$) 162, 164, 167–169
  silicon tetrafluoride (SiF_4) 162, 164, 167–169, 206
  sulfur hexafluoride (SF_6) 163, 165, 206
Etchants (liquid) 3, 4
  buffered oxide etchants (BOE) 26, 41, 42, 52
  piranha 26, 39, 47, 51, 56

Filters 3, 9, 32, 33, 35–37, 39, 70
  HEPA 9, 10–12, 14, 15

INDEX

PTFE 64, 69
PVDF 72
Teflon 35, 37
Fluoropolymer materials 31
  fluoropolymer drums 33
  fluoropolymer containers 41–44

Gallium arsenide (GaAs) 3, 4, 65, 75, 97, 99, 143, 166, 168, 174, 196, 206, 216
Gas analysis
  gas chromatography 69
  infrared spectroscopy 69
  mass spectroscopy 69
Gas purification
  activated carbon 69
  copper 68
  copper oxide 65, 68
  cryogenic adsorption 68
  molecular sieve 65, 67–69
  organometallic polymers 68, 69
  palladium alloy 69
  platinum 68, 69
  point-of-use 65, 68
  potassium hydroxide 69
  sodium (distilled) 69
  titanium granules 65, 68
Gas purity 62, 65

IC modification
  focused ion beams (FIB) 211
  scanning electron microscope (SEM) 211
  scanning ion microscope (SIM) 211
Ion beams
  circuit repair 120
  focused ion beams (FIB) 5, 73, 15, 116, 118, 121, 142
  ion channel 117
  ion projection 117, 122
  masked ion beams (MIB) 115–117
  sources
    field ionization 119
    liquid metal 119, 141–143

Ion exchange 51, 52
Ion implantation 91

Laminar flow 14
  horizontal 10
  vertical 11
Laser doping 207
Laser pantography 218, 219
Laser photolysis 140, 206
Laser projection 218, 220
Laser pyrolysis 140, 141, 206
Lasers
  argon 87, 205–209, 225
  carbon dioxide 203, 204–206, 221
  excimer 74, 76, 77, 84, 103, 204, 205, 217, 220
    ArF 84, 85, 204
    F_2 85
    KrF 85, 86, 106, 204, 210, 217
    XeCl 85, 206
    XeF 85
  HeNe 71, 105, 205, 209, 214
  Nd:YAG 204, 205, 211, 216, 220, 221, 223, 224, 225
  Nd:YLF 210
  Q–switched 204, 210
  solid state 210
  TEA 221
  UV 206, 207
  xenon 211
Lithography 5, 73–125
  contact/proximity 5, 74–76, 78, 127, 128
  e–beam 5, 30, 73–75, 88, 97, 99, 100, 121, 122
    direct write 93, 97, 98
    projection 100
  holography 121
  laser direct write 87
  mix–and–match 5, 74–76, 87, 88, 99
  scanning projection 5, 76, 87, 97, 99, 117
  step–and–repeat (steppers) 5, 73, 75, 76, 78, 79, 83, 84, 86–88, 97, 99, 117, 122, 130, 132, 218

INDEX

[Lithography]
  excimer laser 5, 73, 76, 77, 84–86, 91
  G–line 5, 81, 82, 83, 85, 113
  I–line 5, 81, 82, 85, 86
  step–and–scan 76, 77, 83
  x–ray
    projection aligners 111
    steppers 105, 106, 111–113, 122

Masks 3, 78, 93, 96, 98–100, 121, 130, 132, 134
  materials
    borosilicate glass 84, 128
    quartz 84, 127, 128, 129
    soda lime 84, 127, 128
  mask inspection 5, 97, 138, 142
  mask making 5, 84, 97, 127, 129, 132, 135
    e–beam 93, 96, 99, 111, 120, 129–131, 134–136, 138
    laser pattern generator 134, 136, 137, 205
    optical pattern generator 96, 129, 134, 138
  mask repair 5, 97, 134, 138, 139
    focused ion beam (FIB) 104, 120, 121, 141–144, 210
    laser 140
  phase–shift 6, 82, 132–134
Mean time between failures (MTBF) 20
Mercury (Hg) lamp 76, 79, 87
Metal Oxide Semiconductor (MOS) 1, 22, 49, 108, 127, 185, 193, 208, 216, 225
Metallization
  aluminum 2, 7, 65, 149, 161, 162, 167, 174, 184, 186, 187, 191–193, 196, 211, 217, 218, 220, 224
  aluminum alloys 161, 167, 174, 184–187, 190, 196
  aluminum oxide 162
  copper 162, 167, 174, 220
  gold 220
  multi–level/multilayer 7, 28, 100, 149, 165, 225
  polycide 164
  polysilicon 161, 163, 164, 187, 188, 191, 210, 216
  refractory metal silicides 185, 188
    molybdenum (Mo) 164, 185, 188
    palladium 192
    platinum (Pt) 184, 185, 192
    tantalum (Ta) 164, 188
    titanium (Ti) 164, 185, 191
    tungsten (W) 164, 178, 181, 188, 191
  refractory metals 6, 174, 185, 187, 196
    molybdenum (Mo) 7, 190, 220
    selective tungsten (W) 185, 191, 218
    titanium (Ti) 190
    Ti–W 7, 167, 185–187, 189, 192, 217
    tungsten (W) 7, 178, 190, 191, 210, 216, 220, 225
    tungsten plugs 196
  salacides 196
  silicides 7, 65, 161, 187, 196, 198
  silicon 161, 162, 164, 167, 174, 178, 191
  zinc 220
Microprocessor 2, 51

Numerical aperature (NA) 76, 78, 79, 81, 82, 84, 86, 87, 121, 123, 132, 209

Ohmic/Schottky contacts 184

Packaging 4, 5, 31, 40, 53, 193, 218
Particle counters (air) 9
  continuous flow condensation nucleus 13, 14, 70, 71
  monochromatic laser 13, 212, 213
  optical 13, 70, 212

INDEX

polychromatic white light 13, 212, 213
Particle counters (liquid)
  light scattering 39
  ultra–sound 39
Particle monitoring (point–of–use) 39
Particle removal 33
Particulates 3, 9, 21, 31, 32–36, 39, 45, 54, 57, 70, 71, 91, 147, 151, 166, 193, 212, 214
Photoresist 4, 28, 30, 34, 35, 74, 81, 91, 94, 95, 147, 162, 168, 205, 224
  anti–reflection coatings (ARC) 28, 30, 89, 92
  contrast enhancement 76, 89, 92
  DUV 91
  image reversal 90, 92
  multilayer 6, 117
  negative 27, 88
  novolac 105
  positive 27, 28, 35, 88–90, 92
  post exposure bake 90
  stabilization 91, 92
  stripping 4, 6, 25–28
Physical Vapor Deposition (PVD) 7
Planarization 79, 165, 217, 218
  laser planarization 217, 218
Plasma etching (PE) 4, 56, 91, 152–154, 161, 164
  barrel etchers 145, 146
  batch systems 6, 145, 146, 148–151, 166
  Electron Cyclotron Resonance (ECR) 6, 154
    beam–source 155, 157
    multipolar 156, 157
  helical resonator 158, 160
  helicon whistler 158, 160
  hexode 6, 145–147, 151, 158, 166
  multi–chamber system 149, 150, 151, 155, 166
  single wafer systems 6, 145, 146, 148–151, 166

Plasma stripping 47
Polyethylene containers 40–44

Reactive ion etching (RIE) 117, 145, 152–155, 157, 161, 163, 166, 167, 168
  magnetically enhanced RIE (MER-IE) 6, 158, 159
Read Only Memory (ROM) 1, 2, 49, 51
Random Access Memory (RAM) 1, 10, 32, 51
  Dynamic Random Access Memory (DRAM) 2, 3, 5, 16, 22, 31, 49, 51, 74, 76, 77, 81–83, 106, 107, 112, 113, 132, 193, 210
Reticles 77, 78, 83, 87, 88, 93, 96, 99, 105, 121, 134, 136, 140, 218

Scanning electron microscope (SEM) 121, 142, 211
Scanning ion microscope (SIM) 121, 211
Silicon trench 163, 164
Soldering 205
  laser 220–222
Standard Mechanical Interface (SMIF) 16–18
Statistical process/quality control (SPC/SQC) 53–55
Spectrometers
  atomic absorption 31
  glow discharge mass spectroscopy 187
  inductively coupled plasma 31, 187
  neutron activation analysis 187
Sputtering
  targets 3, 4

Tape automated bonding (TAB) 220

Very High Speed Integrated Circuit (VHSIC) 98–100

234　INDEX

Very Large Scale Integration (VLSI) 1, 5, 9, 16, 25, 28, 57, 70, 97, 129, 146, 163, 164, 165, 181, 184, 185, 192
Via 165, 168, 218

Wafer Air Flow Environment Container (WAFEC) 17, 20
Wafer cleaning
　megasonic 36, 37
　ultrasonic 36, 37
Wafer inspection
　confocal laser 209
　confocal white light 209
　holography 208, 209
　SEM 209
Wafer marking 221, 223
Wafer repair
　laser 210
　FIB 210

Wet processing/etching 3, 6, 25–27, 146

X–ray
　hard 113
　masks 101, 105, 106, 109, 142, 143
　　boron nitride 105, 109, 111
　　silicon carbide 104, 109, 111
　　silicon nitride 104, 109, 111
　resists 6, 114, 115
　soft 101, 105, 113, 115, 123
　sources 101, 104
　　excimer 101, 103
　　Nd:glass 101, 103
　　Nd:glass slab 101, 103
　system types
　　conventional/electron impact 101, 102, 105, 107
　　plasma 101, 102, 113
　　synchrotron/storage ring 101–104, 106–109, 113, 115